Farm

To
Theo and Elaïs
Sue, Ella and Grace

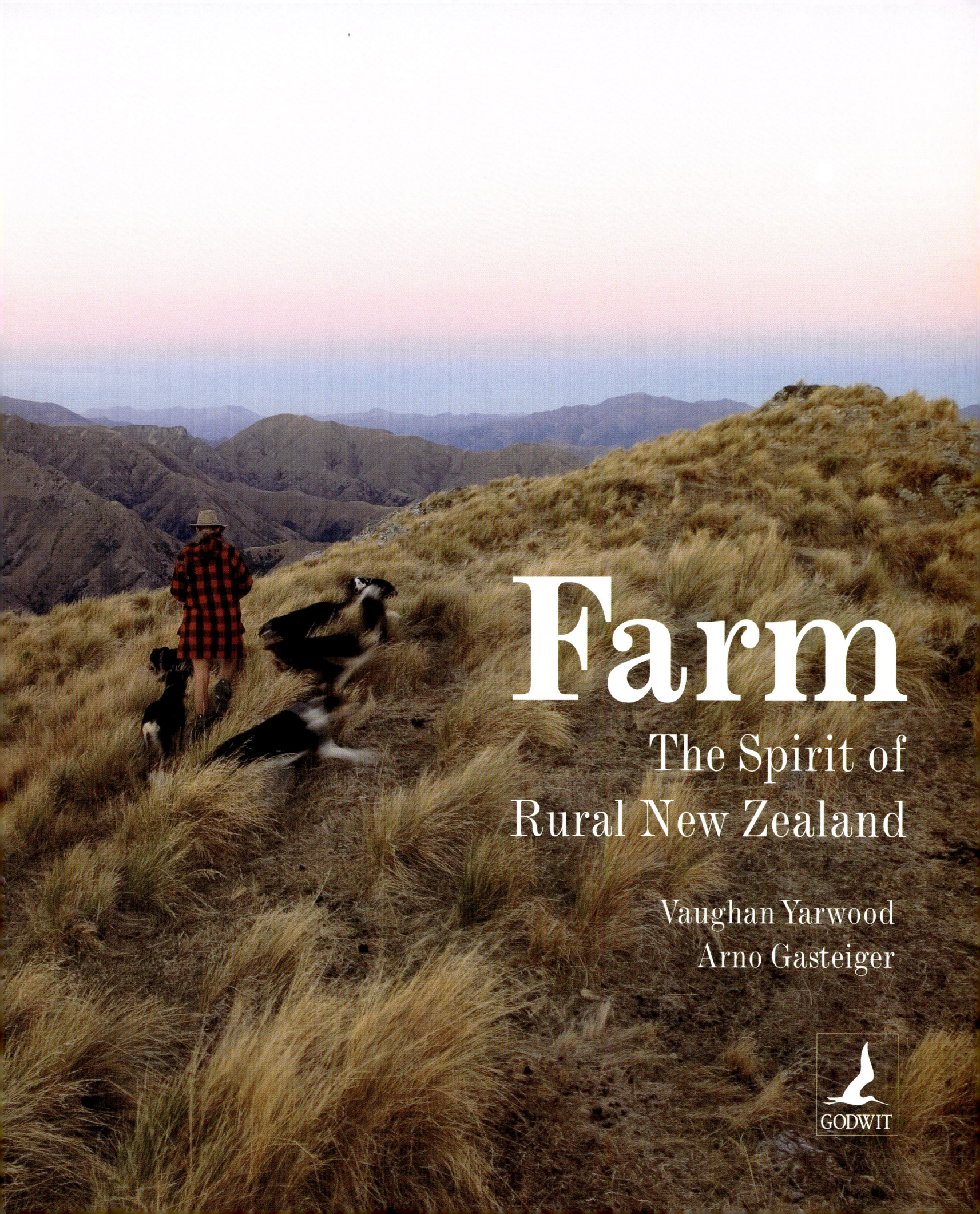

Farm

The Spirit of Rural New Zealand

Vaughan Yarwood

Arno Gasteiger

GODWIT

A catalogue record for this book is available from the National Library of New Zealand.

A GODWIT BOOK
published by
Random House New Zealand
18 Poland Road, Glenfield, Auckland, New Zealand

www.randomhouse.co.nz

First published 2006

© 2006 text Vaughan Yarwood; photographs Arno Gasteiger

The moral rights of the author have been asserted

ISBN-13: 978 1 86962 125 4
ISBN-10: 1 86962 125 5

Cover and text design: Nick Turzynski, redinc. Auckland
Cover photograph: Arno Gasteiger
Printed in China by Everbest Printing Co Ltd

Acknowledgements

A book such as this, which attempts to get behind the agricultural statistics and industry commonplaces and draw close to the spirit of rural New Zealand, would be impossible without the generous support of individual farming families.

We would like to express our thanks to all those who willingly opened their homes and shared their lives with us over the weeks and months that this project was being researched and photographed.

Regan and Margaret Poi at Pakihiroa Station on the East Cape and Peter and Kate St George and Hamish and Bridget Nelson at Nukuhakari Station in the King Country took time out from busy lives to make our stays enjoyable. Steve and Julie Nelson welcomed us to their Waerenga farm while Phil and Jo, Mark and Susannah Guscott at Glen Eden, near Carterton, were unfailingly hospitable and helpful.

In the South Island, Erewhon Station's Christine and Colin Drummond extended a warm high-country welcome that had us feeling like old friends, as did Rodger, Jenny and Guy Slater on their crop farm outside Geraldine.

Andy and Fran Richmond willingly shared the heritage of Richmond Brook Station in Marlborough's Awatere Valley, assisted by Stu and Gin Neal, while in Southland Peter and Norah Thomson at Springburn, and Wendy and Warrick Day at neaby Omokomaru were unstinting hosts.

In all, we travelled more than 11,000 km by car and countless more by plane to seek out compelling stories and to incorporate as broad a range of farming experience as possible. During all this time and through the writing and photo selection process, Nicola Legat at Random House was unfailingly enthusiastic and supportive.

To all of these people and to the many others who supported the project we are endebted.

Vaughan Yarwood
Arno Gasteiger

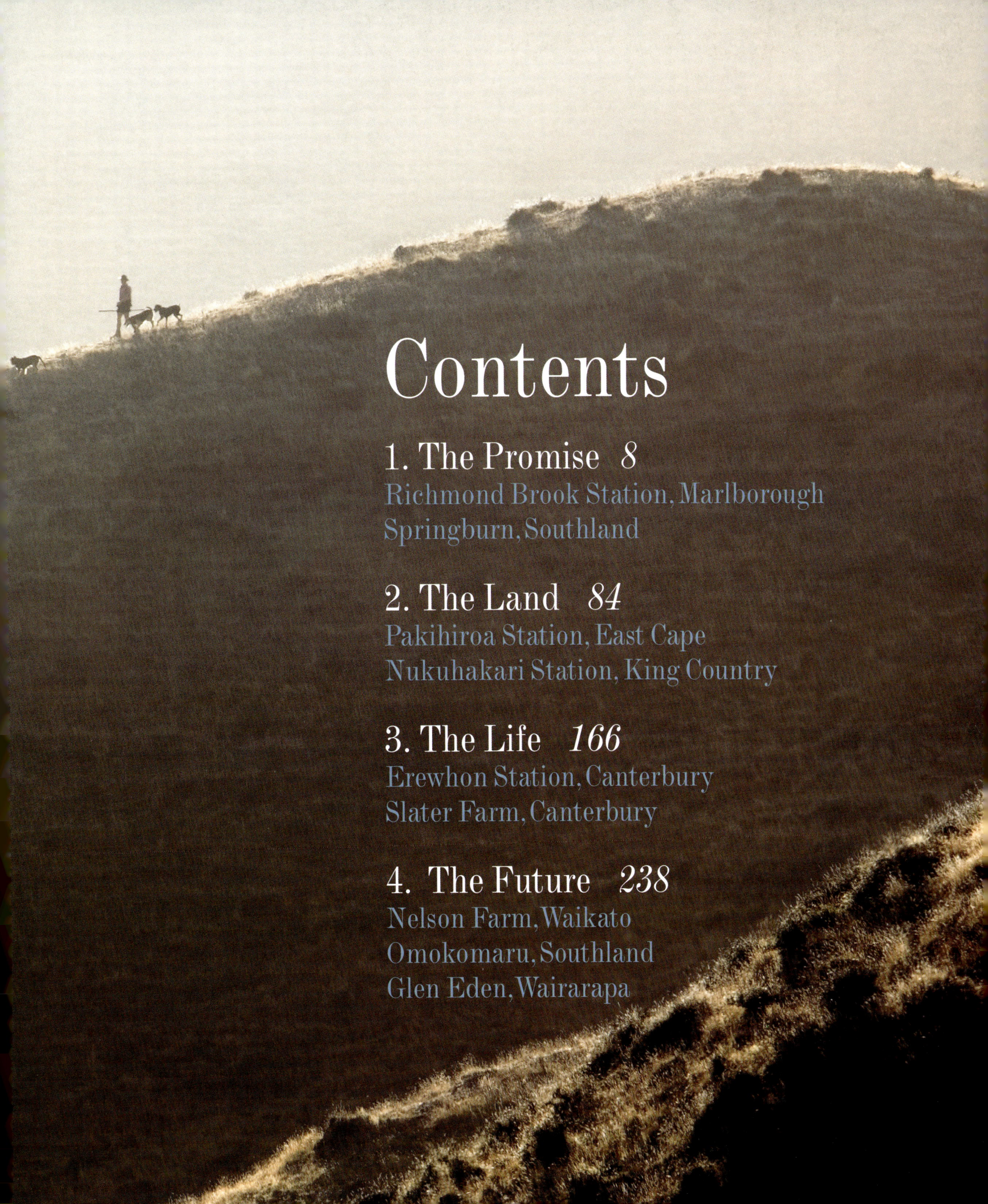

Contents

1. The Promise 8
Richmond Brook Station, Marlborough
Springburn, Southland

2. The Land 84
Pakihiroa Station, East Cape
Nukuhakari Station, King Country

3. The Life 166
Erewhon Station, Canterbury
Slater Farm, Canterbury

4. The Future 238
Nelson Farm, Waikato
Omokomaru, Southland
Glen Eden, Wairarapa

Marlborough was falling under evening shadow as I approached Seddon on State Highway 1 and crested a hill that revealed the squat, domed hills of the lower Awatere Valley. Across the irrigated flats stretched a vast, undulating carpet of vines. At the foot of the hill I swung hard left and followed a side road into the valley, the elevated frost fans and the green geometry of the plantings gradually folding into gathering twilight about me.

It was hard to believe that merino sheep and not Marlborough sauvignon blanc had brought me here. But it was pastoralism that had first put the 'Awa-tree' on the map, and at its heart was the massive two-storey concrete and stone structure that stood at the end of the gravel drive before me — the homestead of Richmond Brook station.

The original station, which took in 14,600 ha of hill country, stretched beyond Blue Mountain to the Ure River and embraced impressive, sun-bathed limestone country. With unsurpassed fine-wool pastures at its doorstep, Richmond Brook became one of New Zealand's most significant merino studs, and in the boom year of 1951 it achieved immortality of sorts with an all-time record price at auction for its wool.

Stations are not immune to change, however. They follow the same patterns of growth and decay as the stock and the crops they raise. They accumulate land and shed it, combine and split, are bought and sold. Their boundaries shift. So it was with Richmond Brook. Around 1900 the back block, at the head of the Ure River, was sold. About the same time another block was carved up for closer settlement. Over the years more followed, including a large holding in 1957. Now the once-sprawling station centres on a home block of 4460 ha — an attractive mix of what used to be called 'wheel-tractor country' and, further back, good tussock land rising to 1000 m.

Despite its shrinking fenceline, Richmond Brook has managed to achieve something rare among South Island stations — for more than one and a half centuries, since Major Matthew Richmond first took it up in 1848, the station has been owned by the founding family. So powerful is this connection that the land has almost become part of the family's DNA.

Now, my knock at the solid, thick-timbered door roused the major's distant descendant, Andrew Richmond, from somewhere deep in the castle-like house. With a genial greeting, Andy swung back the hefty door and led me through a baronial reception hall flanked by a massive fireplace that looked as if it could easily accommodate a spit-roast hog. Gilt-framed oils of Victorian and Edwardian gents hung from the wood-panelled walls, and beyond a brass-faced grandfather clock — whose engraved motto reminded country visitors that 'tempus fugit' — rose a majestic staircase surmounted by well-nourished cherubs bearing lighted globes. It was the sort of establishment from which a man could step of a morning for a day's crutching with a sense that, for all its tribulations, life was no bad thing.

The house was the work of Andy's great-grandfather, Seymour, who had it built as his rural headquarters in 1926. A wall-mounted panel of lights in the kitchen evokes the flavour of those days. The work of the Marlborough Engineering Company, they could be triggered from distant parts of the homestead to summon the domestics. By the time Andy's grandfather moved in, times had changed down on the farm and the maids had gone.

By a happy coincidence, Andy's brother James had now dropped in briefly before heading north for a few days. Arranging to meet the men and women of the land on their land is like herding cats these days, so frequently do business, pleasure or family obligation seem to remove them. I was glad of the chance to talk about a new commercial venture at Richmond Brook that James was steering, and over a beer he put me in the picture.

After a stint of finance work in London and Sydney, James had set himself up a nice little business growing grapes under contract on 57 ha of station land near the homestead. The vines, which at the time of my visit had been in the ground 18 months, were sauvignon blanc — 'the cash cow around Marlborough' — and were growing on grazing ground that farmers of the 20th century would have classed as too good for grapes. In their day the vines, which aren't partial to wet soil, were planted on silty, bony river flats.

The Awatere's clay soils are redeemed from that particular vice by being in the grip of what amounts to a perpetual drought. Almost the first information I had gathered on arrival was that the day's calm weather had come on the heels of three days of howling nor'westers and that the lingering drought had been alleviated not at all by 2.5 mm of rain — 'if you can call that shower of hail we had rain'.

James told me that by contrast Nelson, on the other side of the Richmond Range, was so wet that its viticulturalists had problems keeping their vines disease-free. It was the first of a series of lessons I was to have about the overwhelming effect on farming of the microclimates thrown up by the country's long, thin and geologically contorted islands.

The first crop, some 40 tonnes or so, from Richmond's original planting was due for harvest in six weeks. Although it was looking good, said James, it was a little behind others hereabouts for reasons best known to nature. 'We haven't put our finger on it. There is a bit of altitude up here and it could be that the site is cooler.'

Awatere doesn't crop as highly as nearby Blenheim, said James, but the quality is good. Blenheim soils yield tropical flavours due to the humidity, especially out towards the Marlborough Sounds. The local harvest was more grassy, with a mineral edge. It was claimed that educated palates could even distinguish differences within the Valley.

James thought that maybe he would take another 8 ha and put it in riesling, which is a less valuable grape but a more consistent cropper. He talked of raising the brix levels — the fruit sugar — to improve the wine's taste, of the good 'medium-vigour' soil (high vigour encourages overmuch foliage and puts little into the berries) and of the trade-offs between machine- and hand-harvesting. Growers pay by the vine/kilometre for mechanised picking and by the tonne for manual picking. On small plantations machinery — favoured by James's winery buyer but not, as it happens, by his technical adviser — tends to be less cost-effective.

Establishment costs had been high. Apart from kilometres of fencing and posts, the countless vines, the trickle irrigation lines, compressed air cannons, fertilisers and frost fans, there was the Herculean labour needed to get the vines in the ground — 55,000 in one new block alone — then to wirelift and, later, prune them. Management of the canopy is important and pruning helps let air through and minimises disease.

Sauvignon blanc is a labour-intensive, and therefore expensive, style of wine to grow. If Chile or South Africa achieve the same type and quality as New Zealand, said James, 'we are in trouble'. Nevertheless, the grape has transformed this part of the country. Winemakers are now to be seen driving down Main Street in luxury cars, long-time industry employees have set up as independent consultants, and suppliers of farm machinery have made room in their yards for new equipment designed for the vineyard.

For the farmers themselves, the grape is ambrosia, white gold, a ticket to the good life. Vavasour Wines was the first in the Valley and others soon followed as farmers wrestling to win a living from the bony land woke to find big wineries in a buying mood and ready to hand over three times what the land was worth in pasture. Irrigation schemes thought too wasteful of capital for mere sheep and beef found a ready logic in the economy of the grape, and with the water came yet more hectares of leaf and vine. It has got so that walking the Awatere in the early morning is like entering a war zone, so filled is the air with detonations from the gas-powered bird frighteners and the shotguns of hired hands.

Farmers in the Awatere are harnessing their wagon to an industry that has hit its straps — so much so that it comes as a shock to realise how recent the whole thing is. In the early 1980s the country had just 97 wineries. Now the figure stands at around 500. In the past decade alone, the area under cultivation has risen from 6610 ha to 21,000 ha and grape volume has more than tripled to 166,000 tonnes, yielding some 120 million litres of bottled wine. Export returns likewise climbed from a modest $2 million in 1990 to $302 million in 2004.

Industry cheerleaders emphasise the fact that, despite the stratospheric climb in earnings, New Zealand still accounts for 0.3 per cent of world production. Sober-minded analysts add a note of caution, saying that the global surplus — that is, the difference between world production and consumption — currently stands at six billion litres and rising. That unattractive figure may cause some among New Zealand's more than 600 contract growers to reach for a steadying glass, but it has done little to depress the spirits of the Awatere's new horticulturalists, who have staked much on the quality of the local fruit.

'If the neighbours haven't got vines then they are contemplating vines,' James told me. People were setting up 120 ha to 160 ha blocks, he said, and this when the average size of vineyards nationwide was 11.7 ha. 'These farmers are putting everything on the line. I've never seen such a confident industry.'

James's comment put me in mind of the magnificent optimism, the fevered enthusiasm, that swept through the country's pastoralists in the early years of European settlement — largely on the well-upholstered back of the merino sheep — and which to me has become inseparable from the colonial careers of two unlikely sheepmen: the English satirist and frustrated painter Samuel Butler, and the likeable Swedish pioneer Carl Sjostedt.

Butler's story is perhaps the more instructive. After a failed attempt at becoming ordained to follow in the footsteps of his father, a Nottingham clergyman, the rebellious 24-year-old Cambridge graduate was handed a £1000 annual allowance and despatched to the antipodes to carve out a life for himself as a farmer. He arrived at Port Lyttelton in January 1860 with just one goal — to better himself financially and so win independence from his overbearing father. Once ashore, Butler soon learned that even at that early date — the tender settlement was just 10 years old — all the readily accessible grazing land as far as the Alps had been taken. Further north, the Awatere and Wairau valleys had been considered fully occupied a decade earlier.

'I was amused at dinner by a certain sailor and others, who maintained that the end of the world was likely to arrive shortly; the principal argument appearing to be, that there was no more sheep country to be found in Canterbury,' Butler wrote in a letter home. Undeterred, the colonial greenhorn bought a horse, named Doctor, and announced to disbelieving runholders that he was going 'exploring'.

Butler and his horse clocked up some impressive distances: they reached the Rakaia and the Harper rivers, and pushed up the Hurunui and the Waimakariri, finding nothing. Up the Rangitata. Then beyond Mount Peel as far as the Ben McLeod Range. Still nothing.

In all this roaming, Butler developed little affection for the Canterbury Plains: 'The same interminable tussock, dotted with the same cabbage trees. They are, in clear weather, monotonous and dazzling; in cloudy weather, monotonous and sad.'

Nevertheless, Butler seemed not to know when to quit. He and a companion eventually fetched up in the bed of Forest Creek, which flowed down a narrow valley. They made camp, and set out again next day 'on a clear frosty morning — so frosty that the tea-leaves in our pannikins were frozen, and our outer blankets crisped with frozen dew. We went up a little gorge, as narrow as a street in Genoa, with huge black and dripping precipices overhanging it, so as almost to shut out the light of heaven . . .'

A few hours later they had reached the tops of the Two Thumb Range, and there caught sight of the MacKenzie Plains, an uninterrupted steppe extending southward. Climbing further to get a better view, Butler came up against towering Mount Cook which rose into a cloudless sky. Then, turning, he glimpsed something that no European eyes had seen — an expanse of tussock and snowgrass country, perhaps 6000 ha in size and at an altitude of some 1200 m, nestled between the Two Thumb and Sinclair ranges.

Three months of hard riding had at last paid off. Seizing his chance, Butler made haste back to Christchurch, sketched a location map and applied for the land. What he deprecatingly called 'a useful little run' would become the kernel of a huge holding that he pieced together lower down, between Forest Creek, Bush Stream and the Rangitata River and, classical scholar that he was, named Mesopotamia — 'the land between the rivers'.

Rather than driving a flock into the secret valley straight away, Butler threw up an A-frame hut and overwintered there himself to judge whether it was safe to graze sheep on the tops. It wasn't and he soon moved down to the warmer river flats, shifting in essential supplies — which for Butler included a piano and more books than he could read — and gradually buying out other runholders.

Not everything went to plan. A hut already stood, illegally, on the site Butler chose for his homestead and its large, querulous owner could not be persuaded to leave. Finally the dispute was settled by a theatrical two-day race on horseback to the Land Office in Christchurch. As Butler relates in his autobiographical sketch *A First Year in Canterbury Settlement*, he won narrowly, and once the business was ended, made straight for his solicitor's chambers, where he seated himself at the piano and calmed himself by hammering out Bach fugues for several hours.

> The South Island high country lives and breathes in *Erewhon*. It is as though Butler the colonial pastoralist and explorer had carried about everywhere on his back Butler the writer, the topographer, the daydreaming, notetaking Utopist.

Butler, who enlisted the help of a shepherd, a bullock driver, a housekeeper and a couple of rouseabouts, is said to have disliked sheep. He whiled away the high-country evenings playing his piano and telling ghost stories, and did what he could to satisfy his craving for the stimulant of society by paying frequent visits to his nearest neighbours, 40 km away. He also diverted himself by renewed exploration, this time a thinly disguised attempt to find a way through the alpine barrier to the West Coast. Beyond the headwaters of the Rakaia River he found, but was unable to cross, what would become known as the Whitcombe Pass.

By 1864 Butler was done with sheep and tussock and biting nor'westers. He sold Mesopotamia for $8000, almost doubling his investment, and set sail for England and lasting fame as the author of a caustic autobiographical novel, *The Way of All Flesh*, and of the brilliant satire *Erewhon*.

The South Island high country lives and breathes in *Erewhon*. It is as though Butler the colonial pastoralist and explorer had carried about everywhere on his back Butler the writer, the topographer, the daydreaming, notetaking Utopist. It is obvious from the book's opening pages that, however much he deprecated the 'utterly uncongenial' station life, Butler was never to be entirely free from its spell.

The land in *Erewhon*, Canterbury land thinly veiled, is unsurpassed — 'millions on millions of acres of the most beautifully grassed country in the world' — and the galloping progress of the European settlers has about it an air of New World triumphalism:

Sheep and cattle were introduced and bred with extreme rapidity; men took up their 50,000 or 100,000 acres of country, going inland one behind the other, till in a few years there was not an acre between the sea and the front ranges which was not taken up, and stations either for sheep or cattle were spotted about at intervals of some twenty or thirty miles over the whole country. The front ranges stopped the tide of squatters for some little time; it was thought that there was too much snow upon them for too many months in the year — that the sheep would get lost, the ground being too difficult for shepherding — that the expense of getting wool down to the ship's side would eat up the farmers' profits — and that the grass was too rough and sour for sheep to thrive upon; but one after another determined to try the experiment, and it was wonderful how successfully it turned out.

Butler was, of course, charting the changes he had seen about him as he took up Mesopotamia and laboured to make it pay. He was also revisiting the emotions of youth, first felt in that youthful land:

I am there now, as I write; I fancy that I can see the downs, the huts, the plain, and the river bed — that torrent pathway of desolation, with its distant roar of waters. Oh, wonderful! wonderful! so lonely and so solemn, with the sad grey clouds above, and no sound save a lost lamb bleating upon the mountain-side, as though its little heart were breaking.

Carl Sjostedt — runholder, gentleman and father of a dozen sons — was the flipside of Butler. Charles Suisted, as he came to be known, was a great bear of a man who stood 1.98 metres high in his socks and tipped the scales at 137 kg. On account of his size, it was the Swede's habit to travel with two horses, resting each in turn and rewarding them with a handful of oats from a silk handkerchief which he spread out on the ground.

The son of an ironmaster, he had spent six years with his young English wife in Van Diemen's Land (now Tasmania) working coastal traders and surviving two shipwrecks before quitting the sea for good and going ashore to manage a hotel. An economic slump in the early 1840s took his savings so in 1842, at the age of 32, and with little left but his family and his good name, he forsook Van Diemen's Land and sailed for Wellington. In the small, makeshift town he took up his old profession, soon gaining a reputation as a genial host as proprietor of Wellington's famous Barretts Hotel. Balls, parties and civic receptions at Barretts enlivened the social calendar — indeed, the hotel seemed to constitute the calendar, the guests at its famous gatherings toasting anything and everything with imported wines and fancy liqueurs, dining on oysters from Queen Charlotte Sound and dancing to the band of the 65th Regiment.

Whereas Butler had craved such society, however, Suisted longed for the land — in particular, for land in Otago that he had long had an eye on. In 1847, word from the New Zealand Company that the 'settlement of Otakou was immediately to be proceeded with' had Suisted scrabbling to stake a claim before the next wave of company migrants arrived from England.

Early in 1848 he shipped shingles, building timber and window glass down to Otago, along with some 470 sheep. By year's end he had established himself on the Pleasant River, becoming the first European to take up land in the North Otago runs. His nearest neighbours in that empty country were the Maori and former whalers on the coast at Moeraki, 16 km away. Following the practice of the day, he set up as a squatter, and for the next 10 years grazed sheep, cattle and horses on some 20,000 ha of fine natural pasture northward as far as Oamaru.

Eventually Suisted built a substantial homestead on cliffs overlooking the sea at what became his Goodwood property. It was a long Scandinavian-influenced building with a high stud and elegant double-hung Georgian windows, built from local totara and black pine. Lime for the mortar came from nearby limestone, and bricks were fired from local clay. At Goodwood House the Suisteds carried on their tradition of hospitality, opening their doors to strangers and friends alike.

Suisted's isolated Otepopo outstation was another matter entirely. The shepherds there lived in two-room wattle-and-daub cottages and slept on ships' mattresses under open thatched roofs. Though butter and milk were plentiful, tea wasn't and the shepherds sometimes concocted 'kilmog', or kirimoko, from the stewed leaves of a wild shrub.

When leases became available for purchase, Suisted was not slow to borrow the necessary capital. The Goodwood soil was fertile and readily grew wheat, oats and potatoes — though getting it all to market (and a small market, at that) without trains, steamers or adequate roads was no easy thing. In September 1851, however, Suisted's horizons broadened. That month the first Otago potatoes were shipped to Sydney and within four years the price had increased fourfold, thanks to a crop failure in Tasmania. The next year was less generous. In fact Suisted's potatoes were almost unsaleable — the price fetched did not even cover the cost of landing them across the Tasman. That year, grain exports to Sydney, which had been flourishing, also slid drastically, wheat alone falling to just half its previous value. It was an early introduction to the vagaries of international commodity markets that New Zealand farmers have now been contending with for more than 150 years.

For Suisted and other Otago farmers, salvation came in the form of wool, exports of which doubled in the mid-1850s. Once sent in small lots to Wellington, wool was soon crossing the Tasman aboard purpose-bought ships in 200- to 300-bale consignments for on-selling in London. Despite the good prices fetched, it was a frustrating trade. For one thing, there was the wait of a year or more to learn what price had been achieved in the London auctions. Even buying fresh sheep to stock the runs was often difficult although, thanks to recurring droughts in Australia, they could often be had at reasonable cost. The Australian imports often carried scab, which was laboriously cured with dressings of tobacco. Suisted and a fellow farmer imported 330 kg of tobacco for that purpose in 1852 alone — at a time when Suisted was running almost 3000 sheep. He calculated that the wool clip from a flock of 5000 yielded a profit of $500. He paid his shepherds well — farm labour then as now was hard to come by — and as an incentive to careful management gave them all lambs above a 90 per cent yield.

By the time he sold his runs in 1857 wool prices were soaring to levels they would not reach again until the First World War. Undeterred by the opinion of the *Otago Witness* that 'sheep farming presents visions of quite dazzling wealth', he got rid of everything — stock, buildings, land leases — and packed his children off to Europe and a good education. For Suisted, this was not pretension. Slow land sales had starved the Otago provincial government of the necessary funds for secondary education and the old sea-dog — Swedish baron and a very wealthy man, in the words of one contemporary — wanted to make the right paternal gesture.

Suisted and his wife returned to New Zealand in 1859, though not this time to the land. Instead he refurbished another waterfront hotel in Wellington, comparing it favourably to 'some of the first class hotels in Europe'. Unlike Butler's colonial adventure, however, Suisted's ended badly. Within six months he was bankrupted. Let down by his business partners, in health too poor to make good, and grieving the loss of his wife, he waited in vain for his eldest sons to return from England. They arrived just weeks after he died.

The first of Suisted's old Otago runs were offered for sale in 1861, the year after Butler set up in Mesopotamia. In a fitting twist of fate, the Goodwood house was destined to become a boarding school for boys.

> While in New South Wales, Marsden shipped wool home to England, writing to the governor: 'This will be the beginnings of commerce in the New World. I anticipate immense natural wealth to spring from this commerce in time.'

So, there they are, two figures from the country's crowded pantheon of self-starters; ciphers for the fates and fortunes kicked up by the rollercoaster ride that was colonial agriculture.

Not that agriculture as an industry came ashore with the first Europeans. The sheep Cook liberated at Queen Charlotte Sound in May 1773 died within days and the pioneers who followed, though they kept animals, hardly thought of themselves as farmers.

One exception was the missionary Samuel Marsden, that indefatigable bastion of the church in the South Pacific. Marsden had a more than passing interest in agriculture. While living in New South Wales he had brought Spanish merinos from Cape Colony, running them on a 7000 ha station inland from Sydney, and by 1807 he had a reputation as a talented sheep-breeder. While in New South Wales, Marsden shipped wool home to England, writing to the governor: 'This will be the beginnings of commerce in the New World. I anticipate immense natural wealth to spring from this commerce in time.'

So it was. During the 1840s, thousands of merinos crossed the Tasman to form the backbone of the New Zealand sheep industry. The merino had the advantage of producing superfine wool on poor grazing country, and at a time when local demand for meat was low and refrigeration for export nonexistent, it was the herbivore of choice for runholders. Only gradually did new breeds evolve to challenge the merino's dominance; breeds better suited to wet ground and close handling, or with higher lambing percentages and better meat production. Even so, merino characteristics were often incorporated into the new animals. The country's first, and one of its most successful, indigenous breeds, the Corriedale, was a merino–Lincoln–Leicester cross.

The big runs were on tussock grasslands, which was ploughed where possible and resown in northern hemisphere grasses. Elsewhere the land was burned to encourage new growth. By Butler's day Canterbury was home to a third of the country's sheep, with another third in Otago. Pastoralists didn't entirely ignore the North Island, though, and indeed as early as 1844 Charles Bidwell was driving Australian merinos along the coast to the Wairarapa, manhandling the kicking sheep down the Mukamuka rocks and through surf to safety.

By the late 1870s there were around 13 million sheep in New Zealand on 6000 properties, which extended north through the Wairarapa and into Hawke's Bay, so that they formed a broad, undeviating band that ran northeast from Bluff to Gisborne.

Just how important sheep were to the country's economy can be guessed by contemplating a few stock and trade figures. By 1891 the wool trade was worth $4.1 million, with a further $1.1 million from live sheep and meat exports and some $340,000 from tallow and skins. The all-up total for cattle at that time was $340,000. Given that the grass needed to sustain one cow could feed up to 10 sheep, it is not surprising that sheep were relatively thick on the ground: 17 million in 1891, compared with around 800,000 cattle.

Hence, after wading through reams of such statistics, my visit to one of the homes of the merino — Richmond Brook.

It was quite by accident that, the morning after my arrival, I came upon evidence of the station's pre-eminence in the wool trade. I was standing in the billiard room — a lavish affair with timber-panelled alcoves, an elaborate ceiling cupola and heavy glassed doors opening into a delicately furnished ladies' sitting room (evidence of the house's status as an entertainment Mecca) — when Andy thrust into my hand a framed newspaper clipping. Dated Saturday 10 February 1951, it was an account, in the minute detail beloved by the reporters and editors of those days, of a wool sale. Held in Christchurch's magnificently named Radiant Hall it was, said the paper, the first wool auction since the beginning of the Second World War in which fine wools recaptured the premium they had traditionally commanded.

Apparently it had been a good pastoral season, with less seed and dust in the clip than usual, and the more than 43,000 bales of mostly fleece wool on offer attracted spirited bids from local mills and northern hemisphere buyers alike. The highest price went to a six-bale lot of merino ewe 'from the famous Richmond Brook clip'. At 240 pence per pound, it was a New Zealand record by almost 70 pence.

Andy tapped the glass over that sentence. 'One pound for a pound of wool. Allowing for inflation, I'm told that record still stands,' he said.

Later, we made an impromptu tour of the station's cavernous shearing shed. It had started life as a blade shed with nine stands, but with the coming of electric shears the number of stands had been cut to six. The dark timbers, polished by lanolin and the brush of bodies, were covered with shearers' graffiti — with names, dates and wisecracks that carry a disembodied intimacy.

There were the usual sheep pens and sorting bins for the fleeces, the suspended shearing arms, the races and timbered dividers. In one corner stood an old baling press. With wool at a 30-year low, there was no way Andy was going to part with $16,000 for a new one.

Richmond Brook's merino heyday has now all but ended. As the property has developed over recent years it has become less suited to the idiosyncratic breed.

'They are picky eaters,' Andy told me. 'High country animals, really, and the rougher the country the better. They will eat the tops off and then starve rather than go down into the gullies. They like to be left alone. We usually have to run Corriedales after them to clean up.'

Nor do the difficulties end there. Bring merino wethers (castrated rams) in for shearing and a whole new set of problems arises. They are on flat, fertilised country — neither of which quality do they appreciate. And if they need to be held in the home pastures for a couple of weeks due to rain (since wet sheep can't be shorn) they will stick to the fenceline nearest the hills and start scouring.

It was easy to like the merino.

Four years ago Richmond Brook let the last of its merino wethers go 'to a bloke who wanted them to break-in country'. That is what wethers have tended to be — a pasture-management tool that rewarded husbandry (in years when wool paid) with a yearly fleece. A block would be burned to get new growth started, then hammered into shape with wethers. These days Andy uses cattle to do the hammering and grazes ewes on pastures once picked over by the wethers. The station currently runs 1500 merino ewes and 5500 Corriedales, along with 500 cows and sundry replacement stock — hoggets and heifer calves.

Wanting to get a taste of Marlborough sheepwork, I fronted up early next day for the usual morning meeting at the 'main centre of happenings' — the tractor shed. There I met the operations manager, Stu, and the station's two shepherds, Charlie and Kelly, all of them kitted out in shorts, gaitered boots and work hats and giving every indication of wanting to be up and at it.

Stu, who played in the Marlborough Boys' College First XV alongside Anton Oliver and Leon McDonald, possesses that sort of robust, go-anywhere physique that comes in useful on a farm. He was leaning on a mud-splattered ute and holding forth on the previous day's doings — a rough session fishing out in the Sounds which had even seasoned men thinking thoughts that weren't to do with beer. He was also deriving a great deal of pleasure from the antics of his right-hand man, Scrappy. Scrappy, a Jack Russell terrier, was ripping around the dusty yard trying to get a rise out of the bigger sheepdogs and having trouble being taken seriously.

'Small man's disease,' said Stu with amused forbearance. 'He chases pigs and goats. He's all over the place and he doesn't let up, but when he gets home at night he's tired and grumpy. You don't dare go near him. He's a hard man.'

With the day's work parcelled out, we took a couple of dogs and headed off to bring in two cows for killing. As the song says, there is always blood on the farm, but Stu was not pleased that Lindsay, the contractor, intended doing the home kill in the yards. Other cattle smell the blood and won't go near the place. Still, it was a good thing that he could come at all. The local abattoir had closed the previous year, some said because the supermarkets weren't supporting it, and Lindsay had been busy since October.

Out in the field the eager dogs had difficulty imposing their will. One cow in particular was getting toey. She stood her ground, protecting her calf, and looking as though she might show the frustrated dogs just what a few hundredweight of mother could do.

'That's why we use horses for mustering,' said Stu, striding forward with his arms outstretched. 'You can imagine what would happen among a mob of two to three hundred cows with calves. They show more respect for a man on a horse than for a man with a team of dogs.'

With much coaxing and a few frank exchanges of opinion, the cattle were eventually gated in the yard and we turned to the next job — bringing a mob of sheep down into fresh pasture. It was early still, with pockets of cool air in the gullies and low sun torching the dust kicked up by animals on the move. I knew this was the easiest time on a farm to shift sheep and cattle (parenting instincts aside) but even so the Corries were in three or four minds as to where they should go. With a word Stu despatched a dog across a gully and watched it dart up a spur to reorient a handful of strays.

'In this sort of country a heading dog is important,' he said, eyeing events on the hillside opposite. 'You can always fix a problem with a heading dog.'

Over the coming months I was to see countless evidence of this. Without a sound, one of the keenly intelligent, silky coated dogs would sprint off to a distant part of the action and redeem a situation. Occasionally, a younger, inexperienced dog would disappear over rising ground and lose the plot entirely, or make up its own orders on the trot. But for the most part, they performed enthusiastically, prodigiously and with uncanny nous.

'It's much kinder on stock if you are guiding them with a heading dog, rather than pushing them,' Stu said. 'If you endlessly push them, they get to the point where they won't move without it.'

The sheep were mobbed up now, and Stu brought a big-set huntaway into play. The dog barked assertively, the sheep stirred as one and with a cloud of dust from the dry Awatere earth, man and beast headed down country along the rutted track.

This was the most important time of the year for stations like Richmond Brook. If the ewes flushed well — in other words, put on a good amount of pre-lamb weight — they would take the ram well, scan well and produce more lambs. If done right, flushing could put up to 4 kg on a sheep in just 40 days, which in turn would encourage more eggs to be released and so result in more multiple births.

Richmond Brook was not a fattening farm, so its interest was in lambing percentages. Its Poll Dorsets would be expected to scan at 175 per cent, the Corries at around 160 per cent, and merinos at 135 per cent. The Poll Dorsets were being phased out, however. A previous manager had introduced them with the idea of selling early in the season in order to achieve a premium for lambs before the prices dropped away. Managing three distinct mobs had proved 'a little confusing', though, so it was back to Corries and merinos.

The station focused on sheep breeding because there simply wasn't enough irrigated pasture to fatten lambs. Instead they were sold to finishing farms elsewhere. Caught in its own parched microclimate, the Awatere dictated what farming practice was possible. Old-timers claimed that every eighth autumn was a good one in the Valley, and no one I met felt reason to contradict them.

Flushing on irrigated pasture was a new thing at Richmond Brook, and in truth it was made possible by James's vineyard. The grapes needed water, which meant building a dam up in the hills to collect the flow from a couple of creeks. And if you were going to start shifting earth about, why not make the hole just that much bigger and divert some water for the farm? Which is exactly what they did. The new 300,000 cubic metre dam — that's a lake almost big enough to contemplate water-skiing — holds enough water to irrigate 50 ha of pasture and about the same area in grapes.

The irrigation is Andy's department. Strange to relate, he is not overly partial to sheep. Day-to-day sheepwork is left largely to Stu and the shepherds. Machinery is more in Andy's line, and moving the computer-controlled irrigator about the fields involves one of most versatile pieces of machinery on any farm — the tractor. With its perspex-enclosed cabin and stereo speakers, its myriad sculpted controls and insulated ride, the 21st-century tractor is a galaxy away from the modified old Massey Fergusons that Ed Hillary coaxed to the Pole. If he was that way inclined, Andy could wear a white shirt and tie for his irrigation chores and step down from the cab at the end of it all looking as though he had been through nothing more than a hard day at the office.

Not that the new-fangled machines have it all their own way. A while back Andy's big tractor got stuck, so he stumped off to a flaking shed and fired up the station's old 1948 Field Marshall — a single-cylinder workhorse with spindly front wheels and a take-no-prisoners attitude.

'You can't kill the power on it,' Andy said admiringly. 'It has a 30 hp engine, but so much torque.' Just the thing for hauling some intractable deadweight out of the mire. The outcome of the rescue attempt mirrored that of the 'bigger digger' known to all New Zealand children.

'Came out no trouble,' said Andy. 'Piece of cake.'

I asked whether, given his love of things mechanical, he had contemplated using a helicopter. They were pressed into service elsewhere for the odd piece of farm work and I knew of a high-country farmer near Lindis Pass who did most of his stock work from the seat of one.

As a matter of fact, said Andy, he would love a helicopter for the station. 'I couldn't justify it, but then you don't have to justify everything.'

There have been precedents. Andy's godfather founded Wanaka Helicopters and excused the aerial work on account of the time saving — he had been a farmer himself, with two big Canterbury properties. An image came to mind of Andy (in a white shirt, why not?) piloting a Hughes 300 across the parched earth of Richmond Brook, flying the damselfly-shaped machine over the Depression-era tracks that had been hewn into hillsides with pick and shovel, and tracing the station's outer limits, where in days past boundary riders had laboured ceaselessly to keep stock within its unfenced perimeters.

But enough of fancy.

> The new 300,000 cubic metre dam — that's a lake almost big enough to contemplate waterskiing — holds enough water to irrigate 50 ha of pasture and about the same area in grapes.

I left the dry Awatere shortly after sun-up to a salute of vineyard cannon and was not surprised, just a few kilometres and several low hills south, to encounter a light rain which grew heavier the further along the Kaikoura coast I travelled.

Having experienced the best fine-wool country the land had to offer, I was keen to explore an entirely different farming proposition at the other end of the island — the rolling, emerald-green pastures of Southland, 1200 km away. My journey south took me through a place that has become a symbol for the transforming power of agriculture in New Zealand; a town whose manifest wealth, whose enterprise and bustle, reminded one early visitor of 'the Far East and the tales of old Byzantine times'. It was Oamaru, writer Janet Frame's 'Kingdom by the Sea'.

Anyone driving down the main street — a 40-metre-wide avenue lined with classically inspired stone buildings whose grandeur would not have disgraced Imperial Rome — might be excused for wondering how such lofty gestures came to be made here of all places; how this unquenchable optimism and faith in future prosperity arose.

Oamaru's towering post office, built in the flamboyant French Second Empire style, gives a clue. Completed in 1884, it served a town whose population had quadrupled in just a decade to more than 6000 — making it the seventh largest in the colony — thanks to burgeoning yields of grain, meat and wool on the large inland estates. By then North Otago had consolidated its reputation as one of the country's pre-eminent wheat-growing areas, with more than 32,000 ha of land under the plough and an annual harvest of some 28,000 tonnes (more than a million bushels). The market for grain was boosted by the huge influx of people caught up firstly in the gold rushes and then, a decade later, by the public works and immigration policy of Julius Vogel.

The New Zealand Elevator Company's massive five-storey grain elevator is testimony to the region's productivity. Something of a colonial skyscraper when it was put up in 1883, the building's doors were large enough to admit railway wagons and it incorporated the latest labour-saving American technology. Unfortunately for its owners, the elevator opened as farmers were turning to meat and wool, and the building soon fell into decline.

It was the discovery of gold at Gabriel's Gully in South Otago in May 1861 that first set things in motion, though farmers, slow to realise the economic implications, found reason to complain at the invasion of so many strangers. Soon gold was being discovered further afield, and by the end of the year Otago's population had more than doubled to 30,000. Gold revenue attracted the banks and helped underwrite Oamaru's agricultural infrastructure. Agriculture in turn gave rise to warehouse construction, railroads connecting the town to both Christchurch and Dunedin, and the building of a breakwater to form a safe deep-water port.

Then, in 1882, an experiment was made at Totara Estate, 8 km south of Oamaru, that changed the face of farming for ever. The estate was part of the New Zealand and Australian Land Company's extensive interests, which included 700,000 sheep on 1.2 million ha of land, and it was there that two company men, William Soltau Davidson and Totara manager Thomas Brydone, oversaw the country's first export of meat. Sheep were killed in a newly built slaughterhouse before being despatched by rail to Port Chalmers, where the carcases were frozen aboard the Albion Line's *Dunedin* and shipped to Britain. The trial cargo of mutton and lamb reached London in perfect condition three months later, selling for twice what it would have fetched locally. It heralded the start of what was to become a $5 billion-a year industry. With income from meat as well as wool, small-scale farming at last became viable. Woolstores, grainstores, freezing works, rail, port . . . the whole history of the country's agriculture can be traced in the streets of Oamaru.

At a time when roads to the town and through the Waitaki were little more than rutted tracks through tussockland, the sea was the obvious highway. Yet until the breakwater was built in 1880 to tame the ocean surges and create an artificial harbour in the lee of Cape Wanbrow, Oamaru was an open roadstead and one of the colony's most dangerous ports. Almost everything — from foodstuffs and clothing to news and gossip — came by sea, brought ashore perilously by surfboat. Oamaru's breakwater was its salvation. The sheltered wharves gave pastoralists a conduit for the export of wool, grain and then meat, and so turned the town into one of the so-called 'protein ports' that hard-wired the colony into the greater economy of the British Empire.

My Southland destination, a holding on the flood plain of the Oreti River south of Winton, was one of the farms supplying that protein trade, grazing cross-bred Leicesters for the English market. The land had been taken up in 1867 by John Thomson, an enterprising Scot who had waded ashore through Invercargill mud three years earlier with little but what he stood in — all but two of the family's 14 boxes of belongings having been mislaid.

The voyage out had not been without incident itself — the skipper poisoned his wife aboard ship and was condemned to death. The hangman who was to carry out the sentence himself died at sea while sailing to New Zealand and another had to be found. In all, it took three years for fate to catch up with the skipper.

In that time the irrepressible Thomson had built a sod whare on rent-free land at Winton, becoming what 'would-be witty people', as he termed them, called a 'cockatoo'. There, he grazed cows on the native grasses, before securing a lucrative ploughing contract and turning his hand to fencing and general cartage. With the proceeds of all this labour he bought the 810 ha block that became Winton Plains, in the district now known as Thomson's Crossing. To begin with, before the frozen meat trade and the Leicesters, he successfully cropped wheat, oats and barley there for the Sydney market.

As Thomson's many children married, he carved off slices of the estate for them, so that at his death in 1894 there were no fewer than six homes on the original block. 'The old gentleman is also represented by a large number of grandchildren, making it improbable that the broad acres owned by his sons or daughters will ever have to go seeking for heirs,' the *Southland Times* noted in its fulsome obituary.

Nor did they have to go seeking. I met up with fifth-generation Peter Thomson and his wife Norah on the home farm, Springburn, at Thomson's Crossing. They run 3250 ewes and 800 hogget replacements on 250 ha there and on an 80 ha scattering of leased land. The sheep are Wairere Romney, which give high lambing percentages on the prime lamb-producing country. It was one of those satisfying historical symmetries that the first Romneys had arrived in the South Island the same year as John Thomson.

Peter's father had started out with Romneys then switched to Border Leicesters, which seemed to have higher percentages and to show better mothering ability. In the late 1980s Springburn went to the Coopworth, a Border Leicester–Romney cross. Coopworth lambs carry a little too much fat for the liking of the freezing works, though, so Peter soon went back to a Romney breed. It was a good move. Lamb weights increased, fat decreased and lambing percentages climbed — over the past four years they have averaged 154 per cent. 'It has got to the stage now where we are keeping Romney stock to breed ewe lamb replacements for sale,' Peter told me.

Listening to this sort of talk from sheepmen had me thinking that behind the simple formula of stock units per watered hectare lay a convected and shifting set of equations peculiar to discrete spaces and to moments in time; to the peculiarities of topography, market whims and currency shifts. Here and now, at Thomson's Crossing, it was still worth taking wool off the sheep's back, but there were half the number of Kiwi dollars in the hands of cockies as there were a few years earlier. 'Unfortunately, to get meat we have to have wool, though it involves a lot of work and a lot of costs,' Peter said.

Contemporary New Zealand farms have had to get bigger to survive, but higher stock levels and the shortage of labour has meant a new approach to farm management.

'When I was young, August was the time when my father did the fencing. Now there are jobs like winter-break feeding and vaccinations. This year the two-tooths will get five vaccinations before going to the ram. When I started in 1977 they got none.'

Same with winter feed. Springburn now grows kale and swedes to keep the sheep off grass — if there is a downside to Southland's climate it is poor winter grass growth — and prior to being fed those crops the sheep are put on a strip-grazing pattern. If rain sets in, the animals must be moved more to stop them churning up the pastures. Otherwise the ground will turn to mud and grow less grass in the spring.

One of the practices adopted by Peter and others in the face of these new pressures has been to contract out jobs that they once tackled themselves. Shearing, which is a high-energy, highly skilled and short-lived affair, was a

> Here and now, at Thomson's Crossing, it was still worth taking wool off the sheep's back, but there were half the number of Kiwi dollars in the hands of cockies as there were a few years earlier.

candidate from the first. These days fencing, fertiliser application and even animal dipping have joined the lengthening list of contracted work on some properties. A four-strong dipping gang does the job at Springburn. Using a dual-race trailer equipped with conveyor belt and high-pressure jet, they get through 1500 sheep an hour — a morning's work compared with the three days it used to take Peter and his neighbours when they helped each other out with the dipping. 'It is a saving of time, not money,' Peter said, 'but these days it is not easy to get labour.'

Some things don't change. Back in 1874, old John Thomson was complaining of a district 'heavily weighted with dear labour', though blessed with cheap, fertile land and a good climate. Land fertility wasn't a constant, though, and needed replenishing. In company with others of his generation, Peter's father had taken the blanket approach to fertiliser, habitually dumping 375 kg/ha on the ground every year. Peter did a soil test and found that he had 2250 kg/ha locked up underfoot because not enough lime had been used. By applying lime without phosphate, and so reducing the acidity, Peter increased the grass growth. He now spreads half the amount of fertiliser as in previous years and diverts the savings into minerals — sulphur, zinc and magnesium. 'Someone once said to me that a farmer's primary role was to grow grass and that what he did with it afterwards was up to him. We sometimes lose sight of that fact.'

'Someone once said to me that a farmer's primary role was to grow grass and that what he did with it afterwards was up to him. We sometimes lose sight of that fact.'

With average lamb prices down 23 per cent on last year, Peter recently pulled back a little on fertiliser. Though soil fertility at Springburn had reached a level where he could skip a year entirely, he confessed a reluctance to go that far given the notoriously unpredictable nature of farming. For all he knew, the market for wool might go into free fall in 12 months, or a big chill might savage his lambs.

Even now, things were not going entirely to plan. Peter had some ewe lambs on his hands that he had hoped to sell up north. They were no longer wanted, presumably because of the dry weather. 'My son Pete has just been up there. He tells me that some farmers are killing replacement lambs that they would normally hold on to.' Later, in Canterbury, I would see for myself just how bad things could get when farms are in the grip of a drought.

To make matters worse, the local meat inspectors were on a go-slow to leverage a pay rise, leaving Peter nursing lambs that were getting through the grass needed by ewes for next year's lambing. With the meat works not interested in big lambs at present, the extra kilos the lambs were packing on were not going to translate into a bigger pay cheque.

So, what of the inroads being made by dairy herds in the South Island? How did the numbers stack up? 'My accountant tells me that dairy isn't doing better,' Peter told me. 'The top sheep farmers are making as much, but I look at the people who have started with a few cows. They have bought farms and expanded, whereas sheep farmers haven't. That tells you something. At one stage in the 1990s my brother-in-law was making as much off 80 ha in dairy as I was off 240 ha.'

Having chewed the ends off that particular equation, we got on with things, setting off down one of the Crossing's long straights to move sheep. I was struck by the widespread plantings of flax shelter belts — something I had never seen before. Then I remembered that when John Thomson first took up the property the flood plain had been covered with flax, cutty grass and scrub, and that over the years a great deal of effort had gone into breaking it up and draining it. As late as 1904 there was still a lot of flax about. In that year a Riverton flaxmiller was contracted to cut 500 tonnes 'more or less', most of it growing on an island in a lagoon near the Oreti.

When it came to moving the sheep, Peter had a method that left me slack-jawed. He merely asked them — politely. Pulling off the road, he climbed a fence and opened a gate between paddocks. Then he called the scattered sheep and they came, hesitantly at first, then, filled with confidence — or rather haste — they ran and skipped into the adjoining field. No dogs. No commotion. No striding about the place in an undignified manner.

There was no mystery to it, really. Move sheep enough, as farmers often do at this time of year, and they get used to the routine. It helped that they were just a fenceline away from where the grass was so obviously greener.

Shifting another mob off another lease block down the road gave Peter a chance to employ his four-year-old header dog, Bid, who seemed to relish the chance to show what she could do with recalcitrant ewes.

'She's a good lambing dog,' said Peter, firing up the quad. 'Been doing it since she was eight months old. No one could buy her off me.'

The crack of a shotgun a field or two away had me momentarily back in the Awatere. It turned out to be a groom, grown a bit touchy in the shadow of the event, having a go at a clay trap with the best man.

The sheep behaved themselves, trotting down the road like veterans until they came to Springburn's homestead where, being sheep, they veered as one into the garden. 'This is not good for marital harmony,' Peter mumbled, glancing toward the orchard where Norah could be seen stooping over apples. 'There go the pikelets for morning tea.'

We got the sheep pastured and Peter gave me a tour of the farm's kilometre-long river frontage. All manner of waterfowl — black swans, Canada geese, mallards, teal, spoonbills — were fooling about in the placid, poplar-fringed Oreti. Just below Springburn, on a neighbour's property, lay a gazetted sanctuary. Peter also keeps some of his own property off-limits to shooters to encourage the birds to stay in the area.

Duck shooting, he told me, was 'the' event in Southland. Long-lost sons were known to turn up from distant parts of the country at the start of the season. Peter, though, confessed that for him the camaraderie was the thing. He had grown soft around the edges, he said, these days getting more pleasure out of luring a duck into the decoys than shooting it. As we turned for home he grew lively telling me about sighting a rare visitor to Winton — a marsh crake.

> 'I'm not so naive as to think that floodbanks have saved us. As sure as God makes it rain it's going to beat us,' Peter told me. 'It only takes a little more water for a little longer.'

Norah had obviously chosen not to see the sheep. A country mile of pikelets stood on the stovetop. Perhaps it was the Southland air, but there is something to be said for fresh baking of a mid-morning.

Norah was no stranger to the demands a farm put on a person. She grew up with cows just down the road at Waianawa and her brothers are cow cockies. Up early. In the shed late. No holidays. Now there are sheep in the yard and falling commodity prices.

Her sister married a grain farmer. 'So with cows and sheep and crops we are all busy at different times of the year. It is hard to get the family together.'

Then there are the big events that nature throws into the ring. The Big Freeze of 1996, which burst pipes in the home, froze swedes in the field and hid grass from the sheep for a fortnight. And, being on a flood plain, the water problems. The 'hundred year' flood of 1978, was it? And another in 1980. They came around so fast. Norah remembered water lapping the second step at the back of the house in 1984.

After the flood of '78, which set Invercargill Airport awash, the regional council decided to overhaul the floodbanking. The scheme was still on the drawing board when the 1980 flood came through.

'My father in his wisdom put in a banking system to divert the water outside the flood zone and channel it back into the river system,' said Peter. His own contribution was a way of attaching wire to fenceposts so that any debris washing against it would trigger the release of the wire before major damage was done. Once the waters had gone through, it was the work of a few minutes to reinstate the fence.

Peter and his flood plain neighbours have technology on their side in the form of river gauges in the Oreti 53 km up the road at Lumsden. Accessible by telephone, the gauges warn of an impending flood, giving Peter a 12- to 13-hour window to move stock. Without local rain, a swollen Oreti tends to flatten out by the time it hits Winton, but a reading of 1.5 m above normal at Lumsden, combined with rain at Springburn, is enough to get him out shifting sheep.

'I'm not so naive as to think that floodbanks have saved us. As sure as God makes it rain it's going to beat us,' Peter told me. 'It only takes a little more water for a little longer.'

When we were outside once more, surveying those green Southland pastures as an autumn chill touched the air, Peter made what amounted to a confession. As a younger man, he said, he had no inclination to farm the place. If an opportunity presented itself, he had intended selling and moving to higher ground.

'But this is a good farm, and heritage means more to me now. You learn to live with the river and to expect that the floods will come . . .'

It could have been the voice of Carl Sjostedt shipping window glass south to start a new life on the land, or of Sam Butler astride his horse and with an eye on money-making as the high country slipped under his skin. Or of all those Awatere families with one hand in the woolshed and the other on the vine.

It was, more simply, the voice of the farm.

Musterers head out to bring cattle in from the back blocks at Richmond Brook Station.

Looking westward toward the Awatere Valley from the tops at Richmond Brook reveals the extent of the station.

- Fran Richmond maintains the elaborate garden around the family house at Richmond Brook, an iconic Marlborough homestead.

- The dining hall stands as testimony to the wealth generated by early sheep farmers on their vast estates.

Elegance and durable utility share a home at Richmond Brook. Stained glass bearing the family crest adorns a staircase while an old, well-used fuel bowser continues to fill farm machinery.

The station's restored single-cylinder Field-Marshall tractor still pulls its weight when necessary.

Richmond Brook's new stables catch the setting sun. Horses remain the preferred choice for mustering the station's lower hills.

Gin Neal grooms her horse Hairy, a Clyde-cross, in preparation for a local equestrian event. Shows and the local hunt are an important part of life for Gin and her husband Stu.

Two-year old Angus Neal gets a head start managing animals and farm machinery.

The clean lines of the century-old woodshed embody the pleasing aesthetics and functionality of rural architecture.

▸ A neat 'rug' of crutchings is left by contract shearers on the boards of Richmond Brook's woolshed.

▸ Marginal land in the Awatere Valley finds a profitable new use in viticulture.

▸ Operations manager Stu Neal fetches a carcass from the freezer for his hard-working dogs.

⬆ Shepherd Kelly Drummond enjoys the challenges of working on a new property after the alpine surroundings of her home at Erewhon Station.

➡ A feedout of barley in the dry Malborough pastures gives the sheep a protein flush before lambing.

⬇ A big sky over the Awatere Valley lends spectacle to a day's work in the hills.

A shepherd and his dogs work a ridge to muster cattle for calf weaning.

On the large stations, stock was once controlled by boundary riders. These days kilometres of fences are used instead, but maintaining them is a ceaseless task.

The irrepressible Scrappy shows a gentler side during tea break.

Head shepherd Charlie Simmons musters cattle with the time-honoured tools of the trade.

Farm dogs keenly await orders at the start of a new day at Richmond Brook. With fine wool pastures at its doorstep, the station grew to become one of the country's most important merino studs.

Some of Richmond Brook's 500 Angus–Hereford cattle are brought to lower ground at the end of a successful muster.

For farm-hands, mustering is a highlight of the working year and a chance to explore the less-frequented back country.

A good night's hunt at Richmond Brook helps control the destruction of valuable pasture by pigs. This night, six sows were brought out.

⬆ Romney sheep are shifted at Springburn Estate, on the lush floodplains near Winton in Southland.

➡ The past retains a strong presence at Springburn. William Graham Thomson, pictured in army uniform, built the present house. He shares wall space with 1880 portraits of his parents Julia and Peter.

45   Charles Nanney. To WPP. Thomson.

			£	S	D
Aug 18	1 pak Candles 6d		0	0	6
24	Cash £1	Paid	1	0	0
	2 lbs butter @ 8d			1	4
	Stopped work. 22nd Aug 1903				

Charles Nanney started work here again 18th Nov. at noon. @ 15/-

			£	S	D
Nov. 19,	Balance of Last account debt. 7/-			7	0
21	1 lb. butter @ 8				8
23	1 paket candles 6				6
28	2 lbs butter @ 8 per lb.			1	4
Dec 12	2 lbs butter @ 8 per lb.			1	4
	1½ doz eggs @ 10d per doz.			1	3
19	2 lbs butter @ 8d per lb.			1	4
	1½ doz eggs @ 10d per doz.			1	3
21	1 day off work. 2/6	Paid		2	6
22	1 day off work. 2/6	Settled		2	6
24	Cash £1. 2 lbs butter @ 8		1	1	4
26	1 day off work 2/6 28/12/03			2	6
	Paid				

1904. 5 · 90 lambs of our breed   Ewes 534
        30   620
      620   546
            Robertson Dr J. A. J.   1700

April 24   Robert Robertson 3 bags oats @ 6/   Paid

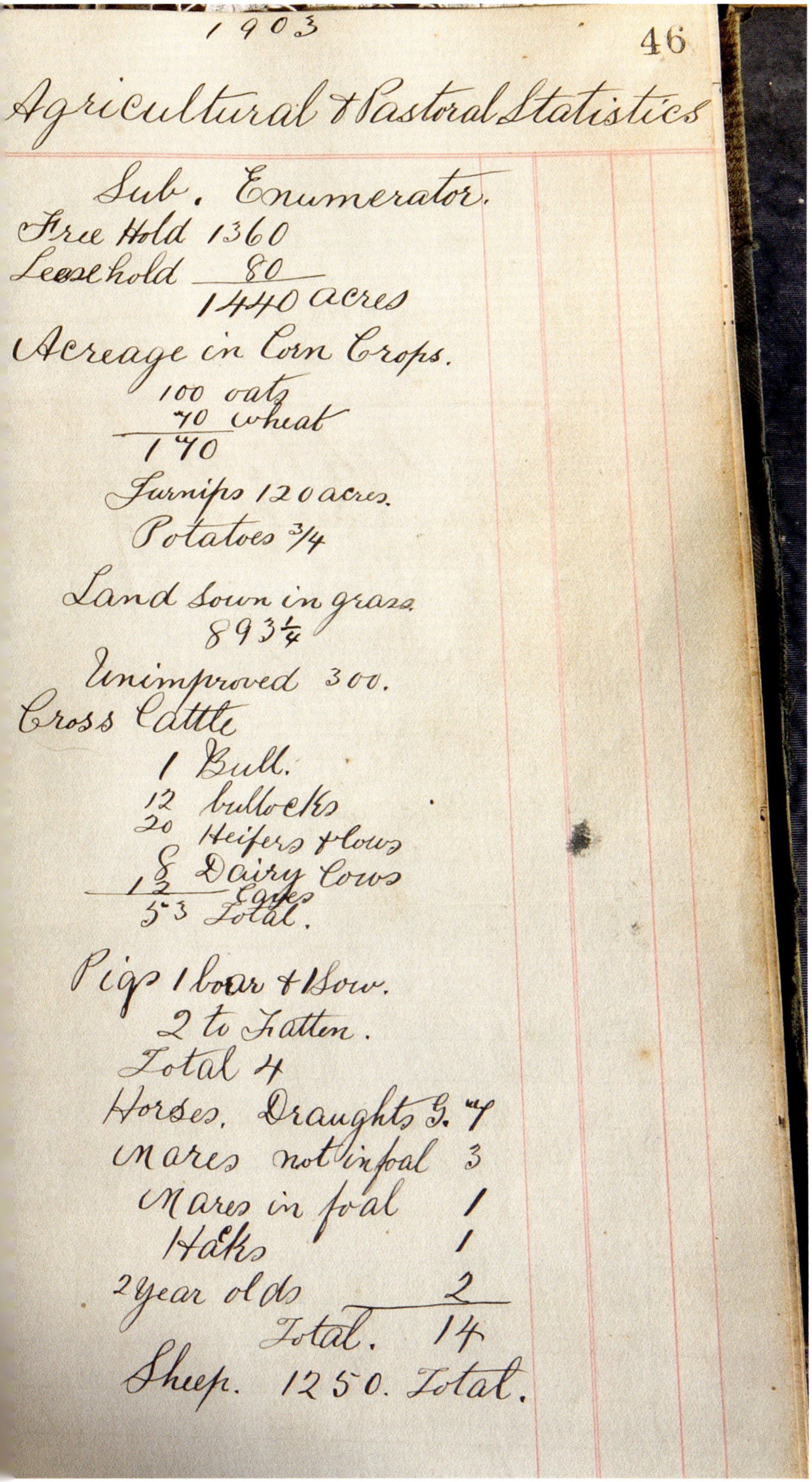

1903

46

Agricultural & Pastoral Statistics

Sub. Enumerator.

Free Hold 1360
Leasehold 80
         1440 acres

Acreage in Corn Crops.
    100 oats
     70 wheat
    170

Turnips 120 acres.
Potatoes ¾

Land sown in grass
    893¼
Unimproved 300.

Cross Cattle
    1 Bull.
   12 bullocks
   20 Heifers & cows
    8 Dairy cows
   12 Calves
   53 Total.

Pigs 1 boar & 1 Sow.
    2 to Fatten.
    Total 4
Horses. Draughts 7
Mares not in foal 3
Mares in foal   1
Hacks   1
2 year olds   2
        Total.   14
Sheep. 1250. Total.

→ Springburn's turn-of-the-century cashbook preserves the daily transactions of a typical Southland farm in Julia Graham Thomson's meticulous hand. After her husband's death, management of the farm became her responsibility.

→ Peter Thomson drafts some of Springburn's more than 3000 Wairere Romney ewes. The first Romneys arrived in the South Island in 1864, the same year as Peter's great-great grandfather.

81 *the* Promise

Regan Poi was not an easy man to get hold of. It took a week or more of trying his number before I was rewarded by hearing his disarmingly warm voice at the other end of the line and instructions on how to get to his front door. Pakihiroa Station had a remote front door. First, I was to take Highway 35 from Opotiki in the Bay of Plenty and follow the spectacular, winding coast road past Cape Runaway and Hicks Bay, through the East Cape settlements of Te Araroa at the Awatere River mouth and Tikitiki on the Waiapu. Then, in the vicinity of the Ngati Porou heartland town of Ruatoria, I was to turn inland off 35 and follow the threading Tapuaeroa River into the mist-shrouded Raukumaras for 25 minutes, give or take.

On that long stretch of road I counted 13 bridges — testimony to the difficulty of getting about in this landscape. The last of them took me from hillsides of plantation forest and across the wide stony bed of the Mangaraukokore Stream to Pakihiroa Station. High up somewhere, partly obscured by trees, homestead lights were burning. As I stopped on the far side of the stream to swing back the gate, two homeward-bound riders emerged from the night and clattered across the wooden bridge, trailed by their dogs. Young fellas from the station, out pighunting. I asked whether they had anything.

'Nah. Not tonight.'

They pointed me in the right direction, then leaned forward in their saddles and coaxed the tired horses up the gravel road. I was to learn that not coming back with the bacon was out of the ordinary at Pakihiroa.

Regan stepped from the house porch and offered his hand. He had only just got back himself from a day out cutting firewood. 'I had to bring back some for the mother-in-law as well, of course,' he explained with an amused smile. He took me indoors, where a wood fire was beginning to drive off the evening chill. Margaret was at the kitchen bench getting food underway. In the next room the girls, Darcel and Arian, and Margaret's niece Davina, were keeping an eye on baby Scarlett and doing their best to chalk up some TV time.

Regan put me right about the phone. He hadn't been absent without leave from the farm all those days. The line had been down. He shrugged. Such things happened in this shifting country. Just as a person wakes to find a hill fallen into a gully or a pile of rubble where a bridge should be, so there are times when there is no power in the plug or dial tone to the phone.

I thought back to the endless repair work I had passed since leaving Opotiki — slumped cliffs being bulldozed from the highway, collapsed culverts resealed, eroded shoulders shored up, the countless orange road-cones becoming a makeshift lei of greeting.

Pakihiroa Station takes in 3145 ha of this troublesome East Cape hill country — 1600 ha 'effective'; the rest in bush. It is a significant piece of Ngati Porou real estate (here I borrow an American expression — tribal elders would flinch at the linking of such a place, such a relationship, with mere coin or title). A realtor would be hard-pressed, though, to find enough flat land on Pakihiroa to make up a modest suburban plot. Mostly it is as steep as a hen's face and just as unforgiving.

The station's southern boundary rises against the flanks of Te Ara ki Hikurangi — a mountain sacred to Ngati Porou. Some say Maui's waka is still up there, in the place where it came to rest. Another big station, Puketoro, lies on the far side but Pakihiroa provides the only road in, the only way to the summit.

Some 2000 to 3000 people a year make the journey to the mountain, many of them to celebrate the New Year. For some the journey takes the form of a pilgrimage — an act of devotion made more palpable by the extraordinary otherworldly carvings that stand at the foot of Hikurangi.

The tourism office in Ruatoria controls who goes up the 1752 m high mountain. The usual practice is for visitors to roll up early in the morning and foot it from the station's carpark to a hut near the base two and a half hours away, then to climb to the summit and walk out the following day.

Visitors caused few problems, said Regan: a badly latched gate that opened when cattle rubbed against it, a car stuck way up the track that had no business being in Land Rover country. Six years ago a party of international students from Waikato University was marooned on the station for a week when the 30 m bridge at the gate was washed out. Regan had been playing rugby in Tauranga at the time. He arrived back the next day and was obliged to ford the swollen river on a borrowed horse, groceries under his arm.

After one and a half bridgeless years, the Department of Conservation, the local council and Pakihiroa itself threw $140,000 on the table to build the present bridge. A few hundred metres away stands a second bridge used by the Poi family, leased from the army by the Malaysian forestry company Ernslaw One. It is surely a sign of living on the edge when the road out to the local shops employs military hardware.

When the station was first taken up, the Mangaraukokore Stream was nothing more than a manageable creek. Then, as the flow grew, fed by increasing run-off from cleared land, a flying fox was set up, to be replaced, as the bed scoured and widened, by a cable car.

When Colin Williams, an early owner, had Pakihiroa, the river flat seemed a good place to build the woolshed. An eight-stand shed, it was one of the biggest around, and became something of a communal resource, being hired out to several farmers in the valley for shearing. Much later, a flood took the shed, along with a haybarn, shearers' quarters, a slaughterhouse and the old, unoccupied homestead.

Regan, who was born and bred just down the road at Tikitiki, was on the station by then, Maori Affairs having taken over the property for a few years before handing it back to Ngati Porou. He worked as a shepherd under his cousin, Parekura (Scarlet) Poi, who years earlier had been an employee of Williams's. When the cousin retired, Regan applied for, and got, the manager's job. He reports to a runanga of the tribe.

Until Pakihiroa got a new woolshed on higher ground, the station's sheep had to be shorn down the road on a neighbouring property. It was hard work, said Regan, because the council had taken the rails off the bridge to give it a better chance of standing up to the debris that floodwaters washed against it.

'Sheep that wouldn't cross the water were run over the bridge and they kept falling off. When we got them to the shed we had to shear the lambs first to give the ewes another day to dry off.'

Margaret pulled a clipping from a pile of magazines and papers and handed it to me. It was a story from the *Gisborne Herald* dated June 2004, with a photograph showing Margaret standing alongside her daughters and children from the Aupouri and Hohapata families out in the wild landscape — striking poses that ranged from stolidly questioning to wearily defiant. Conspicuous by its absence was the Mangapoi bridge — one of the necklace of 13 bridges that wound all the way back to Highway 35. Its destruction by the Mangapoi Stream had condemned the three families to four years of tractor and horse crossings, and they were not about to endure four more. They had had enough, said the paper, of planning their lives around the unpredictable running levels of the river. The gesture stopped short of civil disobedience, but it must have attracted someone's attention. Money was found for a new bridge.

Another page of print caught my eye. It was a story from the local paper, *Whaia te iti kahurangi*, headed 'From Ruatoria to Warsaw'. Again that happening year, 2004. Thirteen students from Ngata Memorial College were off to perform at the International Festival of Folk Art in Poland, among them Regan and Margaret's son Gary. There was a handy list of expressions, first in the language of Ngati Porou, then in Polish. Hello, 'Tena koe', translated as 'Dziekuje'. 'Kare au e mohio' — 'I do not know' — could be rendered 'nie rozumiem'. There were some things to thank the jet age for, not least taking a teenager off a remote North Island station and setting him down slap in the middle of one of Europe's venerable old towns for a couple of weeks.

Then again, perhaps such experiences can have the effect of loosening old ties, old allegiances. At any rate, after the trip Gary, now 19, decided not to stay on the station. The next year he found work in Auckland. Regan, meanwhile, continues to wrestle with the land.

In the morning I ventured out on a quad bike to see for myself the grip that weather and geography exerted on the place. It was a steep-sided, deeply folded landscape, the earth full of stones, the hills bare in patches where the rains had prised off great slabs of topsoil. Some 250 ha of pines and willows had been planted in various parts of the station to help stabilise problem slopes and wet ground, but the eye readily picked out fresh cracks in the shape of huge inverted vees on distant scarps where new slips were building.

It brought to mind a scene from the highly coloured life of Herbert Guthrie-Smith, a 19th-century farmer who, for almost 60 years, had locked horns with the land a couple of hundred kilometres away, at Tutira Station in Hawke's Bay.

In 1931, at the age of 70, Guthrie-Smith had been out shifting sheep — Tutira at one time grazed upward of 30,000 of them — when the Napier earthquake struck. Even as he was being thrown to the ground by the 7.9-magnitude quake, the lanky Scot could not resist speculating on the possible pastoral benefits of the upheaval. Would the tremors revitalise the fields, encouraging cocksfoot, white clover and ryegrass? Might the jolts, in fact, perform 'at no cost to the station the work of a titanic rotary plough'?

Seven years later, what Guthrie-Smith called Hawke's Bay's 'emotional climate' threw down a rainstorm that made the erosion caused half a century on by Cyclone Bola seem almost respectable. On hillsides weakened by the recent quakes the ground appeared to weep mud and saturated subsoils burst out, gouging away turf and creating gaping chasms. Floodwater washed away Lake Ngatapa, which the Napier earthquake had formed, taking with it the Mohaka bridge. Other bridges went too, cutting the road that linked Napier and Wairoa. Lake Tutira rose three metres, and waist-deep silt blanketed 700 ha of the nearby Esk Valley.

> Seven years later, what Guthrie-Smith called Hawke's Bay's 'emotional climate' threw down a rainstorm that made the erosion caused half a century on by Cyclone Bola seem almost respectable.

Years ago, I met a one-time neighbour of Guthrie-Smith, Beatrice Heays — at the time still formidably alert just months short of her 100th birthday — and she recounted the effect of that rainstorm on her station, Te Rangi. Much stock was lost, the telephone line was down for months and, with bridges swept away, supplies were airlifted in. Thinking the road through the gorge would never be rebuilt, she and her husband made the final gesture of surrender to isolation — they jacked the Buick off its wheels.

At the height of the 1938 deluge Guthrie-Smith, a keen student of natural processes, wrote to the Meteorological Office for information on the physical effects of rain. He made a note of the reply: 'For a rate of fall of 5 inches per hour the falling momentum of the rain would cause a pressure of 0.088 tons per acre. The weight of water added per second would be 0.141 tons per acre. One inch of rain weighs 102 tons per acre.'

Tutira's infertile hills, covered in bracken and seamed with gullies and ravines, had broken the fortunes of many farmers since Europeans first tried their hand on the run's 8100 ha in 1873. The early years were times of crippling stock losses, falling wool prices, stubborn debt and the unending battle against fern. Unlike the rich farmland of southern Hawke's Bay, the land bordering the Esk River north of Napier marked the fringes of the agricultural badlands — unproductive country covered in a silty clay and worthless pumice. There the slow work of organic enrichment had been cut short. Its condition prompted Guthrie-Smith to claim that, from a sheep farmer's perspective, New Zealand had been discovered, and Tutira taken up, hundreds of thousands of years too soon. Once covered in forest, it had been burned but not broken in and so languished under a reign of bracken. He had a point.

New Zealand's present shape is less than 10,000 years old and the place gives every indication of continuing to dither over a final form. Some 130 million years of lift and erosion, buckling and tearing, drowning and resurfacing — thanks to its position, straddling the boundaries of two massive, colliding slabs of the earth's crust — have left the country with one of the most varied and spectacular series of landscapes in the world. Plants and animals had barely made a home for themselves in the new land when humans arrived — Maori first, with a desperate need for protein; then Europeans with heads full of northern hemisphere notions and holds packed with Old World creatures and flora.

When the first farmer took the lease, Tutira was an almost unbroken wasteland of fern, and there were not 40 ha of sheep feed on the entire run. South of Taupo, New Zealand's open country was dominated by tussock, but to the north it was the realm of fern. Here, the animals were forced to make their own pasture by a process Guthrie-Smith called 'fern-crushing' — simple enough on good land, but in marginal country full of difficulty. It began with an autumn burn-off of bracken followed by a sowing of grass of whatever quality could be afforded; often little more than floor sweepings. The moment young fern-shoots appeared sheep were driven into the paddock, the number determined by weather and land fertility. If the calculations were right and fate kind, the sheep would subsist on the shoots and keep them in check until the grass took over, permanently banishing the scourge.

But Tutira land was poor and the animals detested the persistent, unpalatable fern that flourished in the wet climate. 'The old sheep died, the young refused to live,' Guthrie-Smith noted bleakly. Fresh sheep were introduced to boost numbers and the process was repeated, the ground, in effect, being 'stamped, jammed, hauled, murdered into grass'. Sheep mustered for shearing were often found to have the wool worn away from stomachs and even flanks through being constantly rubbed against fern and scrub. The fleece of merino wethers, stunted by poor diet, became blackened by sand and grit. It was not uncommon to see the backs of animals drafted in a hot, wet autumn turned green through the sprouting of trapped grass seed.

At Pakihiroa the bracken had been beaten, but beneath the blades of grass lay something less tractable: stones. Where the process of track-making had left banks exposed, the land could be seen in cross section — poor subsoil studded with enough rocks to fill a quarry and capped by a thin layer of topsoil.

'Our fencer sometimes struggles a bit,' Regan said.

I was to meet Pat Boyle, a likeable bloke whose unrelenting tussles with stones had led him to adopt a measured approach to all things. He would start with a standard hole in the turf and soon have on his hands one that was three times the size, thanks to all the rocks he had been obliged to remove. It was enough to test the resolve of any man.

There was a story concerning the visit to Pakihiroa of Tourism Minister Dover Samuels. The minister had been on his way to Mount Hikurangi in a vehicle driven by Selwyn Parata, chair of the station's management committee. As they passed Pat, toiling by the side of the road, the minister shouted for Selwyn to stop.

'I want to shake that man's hand,' he announced, flinging the door open. 'For digging through all those hangi stones.'

'Too much,' Pat said, with a smile and a slow shake of the head, on hearing the old chestnut yet again.

Regan was also amused: 'He probably got a surprise too, to see an Irish fencer on this place.'

Due to the nature of the country — the constant slips, the challenging topography, the stones — grazing management is difficult. Despite Pat's daily heroics out in the field, the fences were not up to scratch. Stock seldom stayed where they were put, which caused mustering headaches and affected pasture production.

'We are always running around with a wire strainer and a bucket of staples,' said Regan. 'There's a lot of patching up to do.'

To make matters worse, some of the old stock trails were washed out and it could take seven hours or more to round up cattle or sheep in the back blocks and get them into a holding paddock before bringing them home.

In late 2005 the station carried out a wild cattle muster — Regan's first at Pakihiroa and enough of an event to attract the attention of a television documentary crew. The Department of Conservation had wanted to cull the animals on its land, but veteran Ruatoria livestock agent and horse breeder Tony Holden had other ideas. He thought a good number of them could be brought out of the bush alive.

The plan was for some station cattle to be brought up to a 600 ha pasture on the station's eastern river boundary as decoys to lure the wild animals out of the bush. Calves were drafted off into the cattle yards to give the decoys a good reason for heading back home once they had done their job.

On a misty November morning six riders saddled up, two helicopters rotored into the air and the muster began. The pilots made light work of the rugged terrain, rounding up the scattered cattle with an agility that went a long way towards explaining New Zealand's outstanding international reputation for aerial stock work — here a nimble sidestep around some obstacle, there a gentle nudge with the skid of a hovering machine against the flank of a reluctant beast.

Over two days some 140 animals — calves, weaners, bulls, cows and heifers — were coaxed and cajoled out of the bush and into the ranks of the usefully employed. Some had been missed in earlier musters, others had come through the ranges. 'I've heard hunters say they've come across cattle in the middle of the bush,' said Regan.

I gained some insight into the difficulties of even ordinary musters at Pakihiroa during a morning spent shadowing Regan and his dogs. Regan was on a quad, but for any number of reasons had a barely disguised dislike of the noisy machines. He told me that a lot of farms at the Cape might go back to horses, especially if lamb prices stayed low and the cost of fuel kept climbing. Even the $400 or $500 he spent every six weeks or so to get seven horses shod didn't sway him toward a mechanical way of thinking.

Toward noon we met up with the station's shepherd Toby who, obscured for the most part by intervening ridgelines, had pulled together a big mob of Romney Perendales. Toby rode over, greeting us with a grin — good humour was part and parcel of Pakihiroa — and trading news, the big man's horse steaming, its hair matted and lathered in sweat.

His boy Joe and cousin Cory rode by, easy in the saddle and ready, with the thirst of youth, for fresh adventure. Before riding on himself, Toby gave out a piece of information that galvanised Regan. 'Saw some pigs down in that gully there.' With raised brows he inclined his head toward a sharply creased streambed choked with bush. 'Big fellas.'

Regan wanted to know when, and whereabouts was that again? And how big exactly? Toby tantalised, parting with shreds of detail in a way Mark Twain would have savoured, before calling his dogs and setting the sheep moving once more.

Lunch and a few necessary chores intervened, but Regan and I were back at that gully soon enough, with suitable dogs and large ideas about the outcome. Regan bent to slip collars off the dogs and to fix a thick leather throat guard to Tahi, an eager pig-dog well marked with scars.

Sensing that a fight was in the air, the dogs grew alert and animated. We picked our way down the hillside, squinting at the bush. Toby sent his header down through it to stir things up, training binoculars on likely spots. Tahi took off into the gully. We ventured out onto a spur to get a view down the creekbed toward the confluence of a bigger stream, the Huitatariki.

Nothing.

'I've got a feeling it won't be long before that paddock is in pines,' said Regan, gesturing toward the cracks up by the road that marked land creep. All around us was evidence of the nocturnal doings of pigs. Everywhere wads of turf lay peeled back where the animals had grubbed the ground, uprooting grass and exposing the tender soil to erosion. The pasture, if it could be called that, had taken a hammering.

Regan got down on his haunches to inspect a fresh print. 'That's how you tell a pig from other animals on the farm. Pigs have these two back hocks.' He set two fingers neatly in the indents. I noticed that some of the prints in the recently dug soil were very small.

> 'They are my boundaries,' said Regan in a tone that hinted at respect. 'Three mountains: Hikurangi, Wharekia,' he turned, 'and Whanakawa.'

It was 4.20 pm and a cold nor'wester had set in. Still no sign of Toby's pigs. We shifted our attention to another part of the station, in the vicinity of a formidable razorback pile called Mount Wharekia. Climbers were sometimes airlifted onto its knife-edged ridge in order to abseil down its sheer western face.

'They are my boundaries,' said Regan in a tone that hinted at respect. 'Three mountains: Hikurangi, Wharekia,' he turned, 'and Whanakawa.'

We found dry ground and sat watching the distant slopes opposite where, high up, a dozen or more pigs had come out of the bush mantle and down maybe 20 m into pasture. Twilight came on and still they hovered, busy about their destructive labour but too high to be easily caught. With horses we might have stood a chance.

Pigs have been blamed for eradicating tuatara on the mainland and for pushing the native orchid and the Chatham Island lily to the edge of extinction. Owners of exotic forests dislike the way they are known to systematically uproot and destroy hectares of saplings, and pastoralists have an understandable antipathy to their factory floor being compromised. There have even been claims that young lambs have been killed and eaten by big sows and boars. Some have said that such unusual behaviour was the result of calcium deficiency, citing as evidence that in the Kaingaroa forest bordering the Ureweras wild pigs had been caught in traps baited with bones.

On the credit side — such as it is — wild pigs, with their liking for fern, may suppress bracken. It has even been suggested that by destroying burrows they help keep rabbits under control.

'From the moment a strange sail bosoms towards a new country, that country's transformation is inevitable,' Guthrie-Smith wrote in *Sorrows and Joys of a New Zealand Naturalist*, one of a series of books that have made him a chronicler of the changing face of New Zealand. He was fortunate to have settled in Hawke's Bay at a time when the effects of European colonisation were accelerating, and for six decades to observe the waves of animal and plant migration that he called 'a later and greater trek of living things'. Alien predators such as stoats and ferrets transformed the landscape with their destruction, displacing native species and forever altering the web of life. Ironically, weasels arrived before rabbits — the cure, as Guthrie-Smith quipped, before the disease — then moved on without trace.

Immigrant birds such as magpies, mynahs and thrushes, recently released in New Zealand cities, found their way to Tutira by various routes according to their habits and needs. New plants appeared and, to Guthrie-Smith's educated eyes, they revealed the unfolding pattern of the country's commerce: Chili grass denoted the blossoming trade with South America, Bermuda grass the transfer of troops from India, Californian stinkweed recorded dealings with the United States, Cape barley the exchange of goods with South Africa.

Meanwhile, local farmers grew anxious to detect the anticipated invasion of rabbits which had been plaguing pastoralists in the South Island and as far north as the Wairarapa since the 1860s.

'There wasn't a sheep-farmer in the province who could not produce rabbit-droppings from his waistcoat pocket,' wrote Guthrie-Smith, 'who would not tremblingly request his friends to smell 'em and affirm they were hares' or lambs' or sheep — anything, in fact, but what they were.'

Given such jitters, it was perhaps not surprising that in 1887 New Zealand's first rabbit board — with the power to levy rates and enforce extermination — was formed in Hawke's Bay. This, despite the fact that the animal flourished far more in the dry south of the country than the humid north.

Once the scale of the country's rabbit problem became apparent, it was hard to find anyone who admitted having had anything to do with their introduction. An exception was Captain Ruck Keene, RN, who had released a dozen rabbits in Nelson in the late 1850s. Twenty years later he was openly lamenting the rash act, saying that the hillsides on his Kaikoura run were now alive with rabbits, that they had eaten the land bare and so condemned his sheep to starve, and that he was a ruined man. He put his losses at the colossal sum of £70,000.

In a bitter irony, Keene had actually sacked two employees all those years ago for shooting at the newly released rabbits. He later told anyone who would listen that he should have rewarded them instead — indeed, that he should have encouraged them to work on their marksmanship.

One of the reasons rabbit boards were slower to get started in the South Island was, perversely, economic. Rabbits were — at least to some enterprising souls — highly profitable. So much so that in Central Otago toward the end of the 19th century professional rabbiters began paying runholders for the privilege of clearing the scourge from their land, the money being made on the skins (which were among the world's best) and, thanks to refrigeration, on rabbit meat.

The figures were staggering. In 1893 alone more than 17 million skins were exported and this rate of killing was kept up, intensified even, until by 1919 the annual total was in excess of 20 million skins. Other figures, though, occupied the thoughts of farmers. They took the form of an equation: $2 \times 3 = 9,000,000$. That astronomical product was the number of offspring a pair of rabbits could theoretically produce in just three years.

The landholders had largely been successful in persuading rabbiters to switch to poisons, forsaking the meat trade in favour of skins alone. Nevertheless, the destruction to land and so to farm incomes continued to be immense. Rabbits, along with the practice of spring burn-offs, had turned much of the tussock land of central Otago to desert. It was only in the late 1940s, with the passing of a law to decommercialise rabbit control by removing any incentive to allow them to breed, that the tide began to turn in favour of farmers.

On Pakihiroa a similar tension was at work. The wild pigs seemingly wandered at will about the place — night riders on the station often encountered mobs of them. The animals were unquestionably destructive and had to be dealt with, yet they provided kai for all manner of gatherings, from weddings and funerals to reunions and school fundraisers. Apart from being impossible in a practical sense, total eradication would carry a social cost.

As the last rays of the dying sun licked the tip of a ridge we called it a day and headed back to the homestead. Margaret and her nephew, Regan Jnr, would be in Ruatoria by now, at their weekly tae kwon do training session. One grade below black belt, they were no strangers to the practice hall or to distant tournaments.

Regan, who had once found himself in the ring, would rather tackle steers. 'I think I must have got on the spirits one night and said yes,' he mused. 'Next day the master stood me on the mat for a fight. That was the first and last time for me.'

Lights were blazing at the shepherds' house. Out front, the lads had a blowtorch to two boars that had been strung up from a beam of the lean-to. They were scraping singed bristles from the blackened carcases with a knife. Dogs sprawled contentedly at their feet. Toby looked pleased.

Could his story of big pigs out in that backblocks gully have been a distracting ruse? The suggestion had the big man laughing delightedly. The pigs had been caught across the river, down on Paul Johnson's land.

'There's a tangi tomorrow out at Waipiro Bay,' Regan told me. 'They'll be going there.'

The tongues of the dogs lolled. The breath that accompanied talk whitened in the cool air. Shadows splayed. It seemed that in the compact pool of light under the stars Pakihiroa found its truest expression — not in the head-count of cattle mustered or the throaty sweep and cloud drift of an aerial topdresser, but in the quiet gestures of community, the shared pleasures of lives fully lived.

On the North Island's opposite coast, and a degree or so of latitude further south, Nukuhakari Station shares with Pakihiroa a sense of remoteness, of occupying a space apart. The station lies on the edge of the King Country, that great 18,000 square kilometre fastness of primeval forest to the west of Lake Taupo.

A one-time sanctuary for dispossessed Waikato Maori, the King Country had been off-limits to Pakeha for a time following the land wars of the 1860s, when it was cordoned off south of the Puniu River by King Tawhiao's aukati, or tapu boundary line. The land's Maori name, Rohepotae, meaning 'the edge of the hat', recalls this line — reputedly, Tawhiao had thrown his hat onto a large map of the North Island to show the land he claimed.

The fate of the King Country was sealed a generation later by the appearance there in June 1883 of surveyor John Rochfort. Said to be so tough that he could work all day without food and still be as fresh at night as when he rose, the 51-year-old Rochfort had arrived to survey a route for the main trunk rail line, a job he carried out with tact and persistence. When it opened in 1908, the Main Trunk Line transformed much of the country it passed through, giving impetus to the settlements of Otorohanga, Te Kuiti and Taumarunui, all once Maori villages.

Nukuhakari Station lies almost 80 km from the grand old track, though, and today the job of getting to it — through the Awakino Gorge out from Te Kuiti, then north along a series of rural roads, sealed and unsealed, and past the wonderfully named Mt Misery — makes its location appear even more marginal.

When I pulled up to the large, modern homestead, Peter and Kate St George wasted little time on formalities. Would I like to head out to Back Bay with them? It being late in the day we took the ute, Peter filling me in on station particulars as we drove.

Nukuhakari stretches itself along 11 km of coastline, with Awakino and Mokau off to the south and the small fishing settlement of Marokopa to the north, just beyond Tapirimoko Point. The station enfolds three river valleys that empty seaward, each defined and separated by high ridges that run to the coast. Back Bay is the northernmost.

Half of the 4050 ha property is in grass. Inland it runs to native bush, which seamlessly abuts rugged, forested conservation land reaching back toward the Herangi Range. Some 600 ha in the middle of Nukuhakari is leased Maori land, brooded over by the 649 m canine peak of Whareorino — one of the station's treasures.

At Back Bay — officially named Nukuhakari Bay after the stream that spills into it — I was surprised to see a distinctly unfarmlike building set on the edge of the grassed field, hard up against a beach of black sand, scattered driftwood and the turbulent, heaving Tasman Sea. It was a small macrocarpa A-frame, a knocked-together safe haven from the pressures of farm work, the timber paid for with goat money and insurance from an old barn that had blown down, the labour snatched from whatever down time the station could spare.

'When we first got here in 1993 we were incredibly busy,' said Kate. 'We built this two years after we arrived. It was lovely. We could come out here for a while and walk on the beach and it felt as though we'd been away.'

Incredibly, when the station was first taken up wool was lightered out from Back Bay through the ocean surges to coastal vessels moored offshore. For years steamers were also the most straightforward way of getting provisions, building materials and anything else to Nukuhakari. The road came through belatedly in 1935.

A few paces from the A-frame, a pot was heating on a wood fire in a rudimentary shelter, the old firehouse having burned down a while back during an animated stag night. The simmering pot was the work of a bloke called Bevan, a station neighbour who was at that moment fully occupied breaking in a couple of horses in a nearby corral. I walked across to take a closer look.

Sporting a broad straw hat and an outrageously full moustache that put me in mind of an Arcadian Albert Schweitzer, Bevan was standing in the centre of the enclosure, holding long reins and persuading a mare to walk around him in stately circles. The younger animal was looking on, displeased and fractious. It could have done, I thought, with a dose of that universal cure-all, Richmond Brook's Scrappy.

> The younger animal was looking on, displeased and fractious. It could have done, I thought, with a dose of that universal cure-all, Richmond Brook's Scrappy.

'Six is a little old for a horse to be broken in. It's easier when they are young,' said Bevan, glancing in our direction. He gave the horses a dispassionate look. 'They are both a little porky. Not a lot of wither on this one.'

'I'm getting too old for breaking horses in,' Kate confided, aware of all the uncomfortable saddle-kilometres ahead. 'Still, it's got to be done.'

'Your quarter horse is good,' said Bevan, walking up to remove the mare's reins. 'What's she like at cantering? If she can't canter she's no good to me for rope work.'

The horse banter see-sawed for a time. A dash of kerosene had been put inside the mare, the 'old girl', to treat an inflammation. Seemed it had worked. Peter offered the opinion that she had done a comprehensive workout. 'Looks as though she's ploughed a couple of acres.'

Then it was time to turn our backs on the corral, the A-frame and a flock of white-fronted terns that had alighted on the sand and clamber back into the ute. As we drove the winding and dipping kilometres back to the homestead, conversation turned to the nature of the station's pasture. Nukuhakari was, it seemed, under a curse of kikuyu grass — an aggressive and unwelcome species that had probably been introduced here to stop the inland advance of sand. Though it failed at that assignment, it did get onto the farm.

As a result of changing weather patterns, the tropical kikuyu was becoming an increasing problem throughout the North Island. The usual cure for it at Nukuhakari was a herbicide spray followed by that farming favourite, the burn-off, and a resowing of ryegrass and clover.

'The only time it is of any use whatsoever is in a dry summer,' said Peter unsympathetically. 'Then it is the only grass that will grow.' He lifted his head to the ridges that towered 260 m or more above the valley floors. 'There are 60 hectares of it up there. Energy levels on the top are as high as ryegrass. It's good for cattle to overwinter on.'

Despite that, and the fact that it harbours fewer of the spores that cause facial eczema — a significant issue in the King Country — than ryegrass, Peter was no friend of *Pennisetum clandestinum* and would be pleased to see it wither and die on the station to the last blade.

Nukuhakari is home to some 9000 ewes and 3000 hoggets, along with 14,000 lambs — 'Romney with a little Perendale through them'. The station was also experimenting with Cheviot rams, which produce hardy hill-country sheep. Some of the property's 800 or so Angus breeding cows were mixing it with sheep, 'grooming' paddocks in a fashion that has become increasingly common on farms. The theory was that fast cropping of the grass in autumn encouraged better winter growth.

Older cattle are also useful in taking worms out of pasture so that lambs and ewes are less susceptible to parasitic infection. Young cattle on the station are drenched every four to five weeks coming into winter and again in spring to counteract the effects of stomach worms until the animals build up an immunity.

'When I was a kid, the drenches were no good,' said Peter. 'Nicotine copper sulphate is what they used. The attrition rate was huge — 20 per cent or more. My uncle would lose 30 to 40 per cent if he was truthful. And sheep were half the size they are now, especially in the King Country.'

In Peter's opinion, the advent of good drenches had as great an impact on farming in New Zealand as did the introduction of superphosphate fertiliser. The arsenal was not huge, though — three families of drenches to fight half a dozen types of worms — and there was always the danger that the parasites would develop resistance to one or more of them. Now, in a new round of biological tit-for-tat, breeders such as Fernleaf Romney Stud at Owhango are selecting for resistance to drench-resistant worms, just as they had earlier bred resistance to facial eczema.

High rainfall and humidity ensure that the King Country is 'reasonably wormy', though not as afflicted as Northland. Temperatures in the winterless north don't fall enough to knock the worms, and to Peter's mind any attempt to run sheep there was going to be like pushing the proverbial substance uphill.

Coming over a low saddle that offered a view of the Mangongu Valley's steeply curving slopes, we were hit by a buffeting breeze. It hadn't been that windy on the station for a while, said Peter. In fact, the weather had been so well behaved and calm that trees had been growing which in any other year would have been knocked back by salt burn. Coastal farms hereabouts struggle in the equinoctial winds that arrive in mid-October and can last for a month. Affected by their proximity to the slow-warming sea, and to the conveyor belt of moist air from the west, they have their own distinct microclimates.

I looked out at the broad grassed plain known as Beach Flat, which ended in a tongue of sand and a deep band of churning foam. To the right the valley made a stepped ascent, culminating in a dizzying lunge for the high ridge. That towering promontory, known as the Lookout, had been the scene, years earlier, of a farming disaster that had ended in mayhem and mass death. Some 600 cattle were grazing up there at the time. Somehow, either by rubbing its nose or licking at the latch, one of the animals managed to open a gate, and the rest went through. Perhaps the gate then blew shut. The cattle walked on, following their noses seaward out along a narrowing shelf of grass on the wrong side of the fence. Push, inevitably, came to shove.

I was to clamber up there myself one morning during my stay to watch the mellow sun laying strips of gold along the valley and see at first hand how precarious and daunting that fatal strip of land was. It was said that the boom of falling animals could be heard from the homestead. In all some 40 cattle were killed that day, and for years a pile of bleached bones lay beneath the bluff. These days all the gates up on the Lookout are double latched.

On the opposite side of the valley lay bones of another sort entirely. In a pitted seam of exposed rock that ran the length of the hillside, slanting shadows marked the caves and hollows where early Maori laid their dead. Peter once called in abseilers to rescue wayward sheep that had survived a fall of 10 or 15 m onto rug-sized shelves of grass among the crags, but he himself has had few occasions to be among the caves. Any human bones discovered up there have been returned to their resting place. The more we talked, the more the valley seemed a vast burial site. Winter storms had even uncovered, then reburied, skeletons in the dunes.

While contemplating the arc of life on the station we were waylaid by a farm worker on a quad who drove up the gravel road past a hillside planted with saplings and braked with a swirl of dust, the dogs at his back straining to regain balance.

'Hamish Nelson, our station manager,' Peter explained. He leaned out of the window and the usual talk ensued, concerning sheep and the shifting of them, the state of pastures and the availability of feed.

Hamish was actually more than the manager, I learned. He was a son-in-law — husband to Peter and Kate's daughter Bridget. The young couple increasingly ran the place these days while Nukuhakari's long-time owners began the process of carving out new lives, lives that still connected with the land they loved.

Kate's thoughts, meanwhile, turned to matters of a more social nature. She was in the hectic final stages of organising a reunion for past and present employees, family, friends and anyone else with an interest in the station. The gathering would, she hoped, celebrate life on Nukuhakari as well as rekindling old memories that could be of use to Adrienne Tatham, who was researching a history of the place. Now living in New Plymouth, Adrienne was the granddaughter of Newton King, a Taranaki business identity and early owner of the property.

The founder of a stock and station company, King bought Nukuhakari in 1909. It was he who chartered the coastal steamer that anchored off Back Bay to offload salt, flour, baking powder and who knew what else through the surf — a year's worth at a time — and to pick up the station's 300 or so bales of wool destined for the Wanganui wharves.

Back at the homestead, while Hamish came and went on farm business and Bridget got food into baby Magnus, I learned something more of the Kings. Newton's brother, Frederic Truby King, had been a well-known physician and the founder of the Plunket Society, an organisation dedicated to the health of women and children. Newton himself was a pioneer of aerial topdressing. With his financial backing, his grandson Miles founded New Plymouth-based Rural Aviation and carried out some of the country's first aerial topdressing from the station. Kate unearthed an old photograph that captured the drama of that youthful, risk-taking industry. It was a shot of a Tiger Moth aircraft on the station's Beach Flat airstrip. The frail biplane had come to rest at a reckless angle, its nose buried in the grass, tailplanes lifted high in the air. One wheel was in a drainage ditch.

To young unemployed Air Force pilots returning home from action in Europe and the Pacific in the 1940s, such accidents must have held little terror. With rehabilitation loans there for the asking and Moths available at bargain prices, it would not have been a hard decision to join the lengthening list of innovators and earn a living above rugged rural fields. The economics were attractive. With an aircraft costing about £300 and a charge-out rate of £13 an hour, a pilot's capital outlay could be recouped in just two long, adrenalin-packed days.

Nor did farmers baulk at the bill. The old method of spreading fertiliser in the King Country and elsewhere had been both tediously drawn-out and exhausting. At Nukuhakari a team of 18 pack horses, each laden with two bags of basic slag, would head out to the station's far reaches where a gang of lads would spread the filthy powder by hand. The men were housed in the shearers' quarters during the weeks that it took to dust the steep, unforgiving blocks of struggling pasture.

In such circumstances, aerial topdressing must have seemed like a form of divine intervention. Rivalry between fliers meant, however, that the willing Moths were pushed to the limits of what a fabric airframe, four wings and a 130-hp engine could take. Anything more than 254 kg of super in the hopper and 30 litres of fuel in the tank would pin the plane to the ground, and pilot fatigue or an urge to squeeze one more run out of a drained fuel tank would often see some part or other of the farm wearing the machine.

Among the Moth's defects was its modest carrying capacity. Pilots compensated for this by spending less time on the ground, fixing spikes to their tail skids to bring them up short on farmers' meagre landing strips and refining their loading techniques to get them airborne again more quickly. One pilot demonstrated what was possible by chalking up an impressive 184 take-offs and landings in a single 11-hour day.

New Zealanders had been fitfully engaged in dropping things from the sky since 1936, when a Hawke's Bay farmer emptied a bag of clover seed over the side of a Moth. It wasn't until 1948, though, that official trials with fertiliser began. The Royal New Zealand Air Force approached the task with style, using an Ohakea-based torpedo bomber to drop around 125 tonnes of super and 12 tonnes of lime on Wairarapa paddocks.

Perhaps the top brass began to feel that such activity was a distraction from the service's principal role as guardian of the skies. Or maybe the government funding for such work appeared inadequate. Whatever the reason, the air force ended its experiments. It was left to Miles King and other entrepreneurial private pilots to seize the day, often with results similar to that documented in Kate's Beach Flat photograph. To give an idea of the scale of such aviation mishaps, in 1952, the year veteran pilot Boyce Barrow first gripped the fertiliser-gummed joystick of a Tiger Moth, there were 148 aircraft at work and 149 notifiable accidents — and that was an improvement on the previous year.

Then came the Fletcher monoplane — a powerful, high-capacity workhorse that enormously improved the economics and safety of aerial topdressing. The Fletcher quickly became one of the most potent symbols of modern New Zealand agriculture and with it an era drew to a close. The Tiger Moth had been a pilot's aircraft, said the aviation romantics. The Fletcher was an accountant's plane.

Whatever the technical advances in the air above King Country farms, however, work on the ground was never easy. Like other properties, Nukuhakari had a huge appetite for casual labour and gave employment to Maori from Taharoa, Marokopa and other nearby settlements, many of whom rode over of a morning to put in a day's work. A 91-year-old neighbour of Kate's confessed that breaking in her own farm in the 1940s had been so exhausting that some nights she went to bed not wanting to wake again.

'If the wives don't want to be on the land, it won't work at all,' said Kate, who last year had Christmas and New Year off for the first time since getting married. 'I had to learn how to put up an electric fence without getting a shock. How to whistle for the dogs . . .'

'You're still not good at that,' quipped Peter.

Kate was not to be put off. 'We don't work a quarter as hard as we once did.'

The King Country was settled late, and by a poorer class of settler. When Kate and Peter began farming 40 km away at Piopio it was regarded as a place to go for cheap land. 'Now it has been tidied up,' said Kate. 'It is a highly regarded area.'

The next morning I saw evidence that the tidying up at Nukuhakari was not at an end. A sizeable chunk of low-lying land in the Ngararahae Valley was being drained and bulldozed for resowing in grass. Wood smoke drifted across the raw, blackened earth and through scattered clumps of trees. The same treatment had been given to 62 ha of swampy ground the year before, with satisfying results. 'We just about turned it into a dairy farm,' said Peter.

I caught up with one of the farm hands, Evan, who was giving reluctant cattle the benefit of Doramectin drench in the station's yards across from the squealing, creaking bulldozer. Having heard that the station had less trouble than many in attracting staff, I wanted to know what people did in their free time.

There was the diving, snorkelling and fishing, of course. Snapper, kahawai and trevally could all be got out on the coast — one favoured spot accessible only by climbing down a rope ladder. None of this was without its risks, though, and there had been several drownings at Nukuhakari over the years including, in 1981, the farm manager.

Eeling was safer. Unlike most places on the coast which had been fished out, the station's streams were alive with eels — thanks in part to Peter's refusal to let commercial eelers in. Then there was pighunting, up in the bush. Or elsewhere. Evan got two on the sand once. Another in a paddock where the steers were now standing.

Peter had offered to show me over the back of the station up toward the bushline in the afternoon, and true to his word, he fronted up on a quad. First we took a closer look at the coast, stopping to walk over a splendidly placed meadow on the cliffs with fine views north and south along the spray-misted shore. Somewhere here, perhaps a few metres that way, was where he and Kate would build their retirement home.

I was struck by the freedom of that gesture — to take a gander along 11 km of pristine sea margin and say 'Here', when one could equally say 'There, on that distant hazy point', or choose any place in between. To find a site by walking the land at will rather than by scanning an agent's window seemed to reconnect with an earlier, more ample time.

> To find a site by walking the land at will rather than by scanning an agent's window seemed to reconnect with an earlier, more ample time.

At Taungaururoa Point, electric fences had been strung along a walkway to keep cattle from an old pa site, one of more than 18 that have been recorded on the station's coastal spurs and ridges. The local hapu say there was a pa on every peak, a kainga in each valley. There are people still alive who remember seeing on nearby Tirua Point the remains of an old Maori whare, now claimed by decay, in which the Ngati Toa chief Te Rauparaha once hid. The seasoned warrior had been leading his people on a long and dangerous migration from their ancestral home at Kawhia to the safety of Kapiti Island.

Peter told me he would probably stay on and put himself in charge of planting up the coastal land with native plants and rolling out electric fences to deter stock. Not much could be done about the kumara pits, which were badly eroded when he arrived, he said, but the flax was coming back nicely.

Quitting the coast with some reluctance, we made for the grassed hills that rose toward the thickly bushed slopes of Whareorino. The blackish Egmont ash that lay under the station's coastal blocks here gave way to reddish Mairoa ash. Once through a series of gates, we found ourselves in a lush pasture, cheek-by-jowl with the native forest that marked the sprawling 'unimproved' half of Nukuhakari Station. Peter turned the key on the quad and farm quiet flooded over us. The few cattle nearby fixed us with bovine stares. Ears twitched.

'It's hard to keep up with the changes that have happened over the past 30 years,' Peter said, eyeing the state of this patch of ground automatically. 'There have been more advances in farming than in any other industry.'

It was a good way of life, as well as a business, of course, but 10 or 15 years ago there had been a big shake-up. 'Forget about it as a way of life then — it was survival.'

Recently things had come right. The cycle had moved on. 'The past five years have been the best of my life. Other sheep farmers say the same. Nothing to do with the government. We have just struck a good pattern.'

I raised the unsavoury subject of wool prices. Yes, Peter admitted, they were shocking and getting worse. Farmers were almost at the point of reverting to 'full wool' sheep — allowing animals to carry 12 months of wool before shearing. Normally in the King Country shearing is done twice a year but, thanks to low prices and the high cost of shearing, said Peter, some farmers had now adopted an eight-month cycle — shearing three times every two years.

During this conversation, our eyes had increasingly drifted toward the peak in front of us. Peter began describing the climb, and some of the people who had accompanied him on it, with what seemed the beginnings of wistfulness. I suggested that climbing it now might be a good idea. That was all the cue he needed. Our way was any-old-how, straight up through an untracked tangle of creepers and ferns, crossing tumbling creeks and hauling ourselves around rocks as best we could. Farm gumboots gripped the slippery ground as badly as any other boots of ascension might be expected to do in such circumstances, but after a good while we emerged into a clearing and found ourselves with nothing left to climb.

We stepped along a rocky, lichen-festooned path that gently rose to a slab of stone bearing a trig. There were views east and south across the forested expanse of the King Country, and in the opposite direction, of the station and the land lying toward Marokopa.

Peter talked of opening the station to trekkers. Of cutting trails through the bush and building two or three sleep-outs. That was one of the reasons they had bought here in the first place — as a possible tourism venture for the children if they didn't find jobs. They did, though. Bridget for years worked as a district nurse. Nick and Anna ran businesses in Auckland.

We stood beside Whareorino's trig and squinted into the distance.

'I wish I could have my life all over again,' Peter said at last. 'I love it here.'

There is a sign near the gate at Pakihiroa Station on the East Cape that could equally stand at Nukuhakari, or at any farm in the country. It would sit well, too, on that old trig of Peter's.
Toite Te Whenua, it reads.
The Land Remains Forever.

105 *the* Land

Otherwordly sculptures, carved to mark the millennium, stand on sacred ground at the foot of Mt Hikurangi on the East Cape. Pakihiroa Station, owned by Ngati Porou, forms the only access to the mountain.

Station manager Regan Poi is a privileged guardian of the powerful carvings, which were created in Gisborne under the supervision of master carver Derek Lardelli.

The head shepherd's house at Pakihiroa Station looks east over the Tapuaeroa River. The restless river greatly influences life in the valley, sweeping away bridges and hindering access.

⬆ Margaret Poi, who raises her family on Pakihiroa, fought successfully to have a bridge on the access road rebuilt after years of troublesome tractor and horse crossings.

➡ Joe Hohapata, 16, has chosen to follow in the footsteps of his father Toby, Pakihiroa's head shepherd.

⬇ A modern bridge makes catching the school bus easier for children at the station.

Erosion plays havoc with farming on the East Cape. Much time is spent repairing damage and stabilising hillsides and riverbanks.

Sheep are drafted in the station's yards for trucking to the freezing works. Others are destined for the station's run-off farm further down the valley.

Pakihiroa Station is run by a runanga of Ngati Porou on behalf of its 65,000 shareholders. Regan Poi (far right) manages the property as an efficient business helped by others including head shepherd Toby Hohapata (centre).

↑ The station's new yards, on higher ground, replace the original yards which, along with the woolshed and other farm buildings on the river flat, were destroyed by the floodwaters of the Tapuaeroa River.

← Full-time fencer Pat Boyle faces the neverending job of repair work, including the rails at his back, damaged during a muster of wild cattle.

↓ Regan prefers his horse to a quad bike when heading into the unforgiving terrain of the station's backblocks.

⬆ Craggy Mt Wharekia forms an imposing backdrop to a mob of ewe hoggets on the move at Pakihirou Station.

➡ Regan Jr warms up outdoors before heading off to Ruatoria for his weekly tae kwon do training. Two years ago his cousin Gary went further, travelling with a school cultural group to Warsaw.

Farm hand Hemi Reuben displays the spontaneous humour that enlivens work on the station.

Mt Wharekia (left) and Mt Aorangi, bordering Pakihiroa Station, exert a strong presence, as does the land throughout the East Cape.

Sea breezes ruffle pasture along Nukuhakari Station's 11 km coastline, catching the attention of Rock, one of the station's heading dogs.

The King Country station carries many reminders of its distant past, including evidence of early Maori settlement here at Back Bay.

Coastal influences are readily apparent in home life on Nukuhakari, where Kate St George maintains an immaculate lawn within sight of the sea.

Long-time owner Peter St George has handed care of Nukuhakari Station to his son-in-law Hamish, but when called on still lends a hand, as here, sorting inferior cattle for sale.

⬆ The 8 km unsealed access road calls on the driving skills of Lawton Job, whose truck and trailer unit hauls excess stock from the station before winter sets in. ➡

⇒ Pushed inland by winter storms, sand dunes on the move claim pasture in the farm's coastal valleys.

⇓ A modest A-frame, built with farm labour on the far side of the station, offers time-out on a pristine coast

- Station manager Hamish Nelson oversees an efficient, productive business. Like many contemporary farmers, he is part of a discussion group that meets regularly to swap information and share new ideas.

- Nukuhakari Station was a pioneer in aerial topdressing, which soon replaced arduous hand-spreading of fertiliser by toiling work gangs. The new generation Hamilton-built Cresco aircraft can carry a 1.8 tonne load and this year dropped 620 tonnes of fertiliser on the station's pastures.

151 *the* Land

153 the Land

Kikuyu grass, introduced to stabilise coastal dunes, is now widespread on the property and in summer gives the pastures a lush appearance.

Farm hand Evan Kete enjoys the lifestyle on Nukuhakari. Offering the chance to fish, dive and hunt, the station has little difficulty attracting staff.

Inland, the station runs to native bush which cloaks the 649 m peak of Whareorino, near forested conservation land. There are plans to cut trails for guided hiking tours.

◀ Effective drenches have transformed farming in New Zealand, though there is a continued risk of worms developing drench resistance. Nukuhakari has two yards to reduce stock movement across the sprawling 4050 ha property.

The station's Angus breeding cows are used to 'groom' pasture. Heavy grazing improves winter grass growth and helps remove parasites that are harmful to lambs.

The Tasman Sea breaks against the black sand fringing Nukuhakari. Seen at dusk, the station's high coastal ridges afford captivating views beyond its southern boundary to distant Taranaki.

163 *the* Land

Sheep tracks sculpt the unusual formations of the station in patterns reminiscent of rice terraces.

Eddie picked his way over the river stones, grey-flecked and laced with white, which shifted and concussed hollowly beneath his broad hooves. I turned in the saddle to look back up the Clyde River toward the Southern Alps which, beyond Perth Col, cradled the Lambert Glacier and the Garden of Allah. The broad shingle-bed itself, here some two and a half kilometres wide, lay between two looming ranges — the Potts and Cloudy Peak.

This side of the ice-crusted peaks, the waters of the Lawrence quit their narrow gorge to join the Clyde, which in turn fed the Rangitata. Further down, on the opposite bank, the Havelock did the same with run-off from the Two Thumb Range. To the west and south rose the sharp, chiselled ridges of Sam Butler's Mesopotamia. It was an endlessly unfolding landscape of mountains and scree slopes, given separation and scale by the network of broad, stony river plains. Just to stand in such a place was to feel the cool breath of freedom. In every direction lay a storehouse of new vistas and fresh possibilities.

    I reined in Eddie so that I could savour the silence, the bigness, of Erewhon Station. It was more than half an hour since I had taken my leave of Christine (Chrissie) Drummond, one of the station's owners. She was now beyond sight up the Lawrence with a packhorse, leading an English visitor to one of the empty mustering huts. Earlier, I had watched her load repair gear for the hut, and enough meat, potatoes and other such tucker to keep half a dozen men happy. The big Clydesdale under me took a step and twitched his ears. With a last lingering glance upriver I clicked him on, toward home.

Musterer Perry May had been the first to attempt to put words to the peculiar poetry of Erewhon for me. A son of North Canterbury, he had set his boots against the sides of steep country the length of the South Island in the pursuit of his vocation. 'It would be the most rugged place I've ever mustered on, and that's a fact,' he told me over a brew of strong tea. 'It's a great test of a person's character. If you come here and enjoy it, and want to come back, you could class yourself as a musterer.'

Perry began mustering at 14, and has walked the beat at some of the country's iconic runs, including sprawling Molesworth Station in Marlborough, 70 km from Richmond Brook at the headwaters of the Awatere River. In the 1960s he was part of a gang that mustered on four stations in the Awatere — at the close of the era when musterers also stood in as shearers. Back then 70,000 lambs came out of the Awatere basin in the first two months of the year. These days there are just a tenth that number.

'Erewhon is the same as it was a hundred years ago, that's the beauty of it,' Perry said. 'Sometimes you'll get stuck on a bluff somewhere and it's a thousand feet straight down, and you think, what am I doing here? But the next day you are back.'

Chrissie was at the table sharing the brew the night we talked. She agreed with Perry. 'It pinches your heart, this country.' She confessed that when she first came to Erewhon she found river crossings with sheep difficult. 'You get a couple of good swims out of them, then that's it. They get heavy with water and become hard to move. There were times when I just cried with frustration.' More than once she has resorted to pulling sheep across with a rope tied to her horse's tail.

It was with good reason that Mona Anderson called the published recollections of her days at Mount Algidus Station *A River Rules My Life*. Mona wrestled with similar topography; Mount Algidus lies on a spur of the Southern Alps at the head of the Rakaia Gorge, 50 km northeast of Erewhon. For access to the road leading down to the plains, the station was at the mercy of the Wilberforce River, which feeds into the many-stranded Rakaia 10 km downstream.

One way or another, rivers dictate terms of life at Erewhon and other high country runs just as surely as they do for Regan Poi at Pakihiroa on the East Cape. The Clyde, that broad, rock-strewn highway to the back blocks of Erewhon, still determines the way the station's owners go about mustering. At the time of the autumn muster, when like as not the Clyde flows strong and full, the river often becomes impassable to utes and farm bikes. Nor would bulldozing a road through offer a solution. With so much shingle on the move, just keeping a road to the back of the station open would be a full-time occupation.

The time-honoured answer is to do what Chrissie and her husband Colin do — hitch half a dozen Clydies to a big-tyred wagon, load it with hardy blokes, sheepdogs and provisions, and set off in the direction of the Lawrence and points north — often in the teeth of an icy, rain-peppered wind. At the close of day, the mustering gang then sets up in a hut or under canvas, while the horses find pasture, the dogs are fed and the cook fires up the camp oven. Next morning, while the lads set off on foot up the rugged gullies in search of sheep, Chrissie packs up the gear and heads off with spare horses in tow for the next camp.

The autumn muster involves bringing 1500 or so merino wethers down from the rugged back country before the onset of winter. It is a process lasting up to two weeks that Colin calls 'as much mountaineering and exploration as mustering'. In the early days — the Drummonds have clocked up eight years at Erewhon — Chrissie learned to break-in horses, train sheepdogs and cook on an open fire. One year they took everything for the camps in on five packhorses. The horses disagreed with the arrangements and took themselves off home. 'We called it Camp Runaway,' said Chrissie. 'It was a long walk back.'

A newcomer might incline to the view that using a chopper would offer musterers a technological leg-up out here and justify its cost in time saved. Perry thought not. No stranger to helicopters, he saw no role for them on an Erewhon muster. 'They have their uses, but when you fly you get a different perspective on the country. If the tops are closed in with fog and it clears at two o'clock you can fly out to do some mustering — and you can get caught out. Going in that way also limits the number of dogs you can take.' He shook his head. 'In this country, I don't think it would work.'

Perry speaks from experience — he lost a good dog once through airlifting. A shortage of space in the Jet Ranger limited him to just two heading dogs and two huntaways — he normally works with eight. The day's muster was long and hot and though another musterer arrived to carry on the work, Perry's few dogs had laboured too hard on the baking slopes — one died of heat exhaustion.

So, at Erewhon the old ways prevail. Colin first met Chrissie at Clayton Station when he arrived as a casual musterer to find her in the gang. They soon discovered that they shared a love of the old ways of station life. Today Erewhon is perhaps the only station where musterers still walk in to their beats — and that suits the Drummonds just fine.

A week or two before I arrived, Chrissie had taken a 50-strong herd of cattle to the broken back country, boundary riding for a spell to keep them there. She and Colin had also put ewes out there — dry weather had knocked back the pasture at the front of the station.

It doesn't take a very long conversation with Chrissie to make you realise that any reason to saddle up and head up the Clyde has got to be a good one, and that if no ready-made excuse is to hand, one can be got soon enough. Perry put a name to it: Gully Fever.

'I'm touched by it,' he admitted readily enough. 'I've never had a feeling as good as taking a young dog out onto a hill and working sheep.' It was no throwaway line. Perry had crammed a lot of experience into his 58 years, ranging from shearing to flying helicopters. 'A fear of mine is looking at a hill and knowing bloody well I can't get up there. I don't know how I'm going to handle that.'

Being one of the fittest-looking men I have known, that particular event seemed a long way off for Perry, though I could see in his frank gaze that it constitutes something of an undercurrent of lingering dread.

Riding back from the Lawrence, I had ample opportunity to study the hills and the way the weather played about them. Chrissie had told me that Erewhon didn't often get southerly winds, unless they blew hard, because the tail end of the northerlies higher up tended to block them. 'These gorges are the home of the nor'wester,' she said.

With all rain in these parts coming out of the northwest, the station was usually what she called 'summer-safe'. However, the winter of 2005 was too good as far as Erewhon was concerned. Not much snow fell and spring came early. Rainfall was minimal and the year ended as the second-driest on record.

That was bad enough, but I thought about what could go wrong when the weather dial turned the other way, as it did in the great white-out of 1867. That harsh snowfall had blanketed much of the South Island and caused grief to farmers from Marlborough to Otago.

I had read the letters of that hardy, Jamaican-born settler, Anne Stewart — known to the world as Lady Barker — in which she recorded how it affected her at Broomielaw Station in the upper South Island, on the southern banks of Selwyn River.

At the end of July, having being delayed by unseasonable rain for a fortnight, Lady Barker's husband left to fetch flour, coal and other badly needed supplies from Christchurch, in the dray. That same day several visitors arrived and talk turned to the dismal, misty weather and the odd procession of ewes and lambs arriving unbidden on the homestead flat.

Next morning dawned with a steady fall of snow and after a meagre breakfast — those provisions really were needed — the company passed the time as best they could. To everyone's surprise the following day brought no improvement, and snow now covered the fowlhouse and pigsties and lay two metres deep on the verandahs. At dinner the six beleaguered souls sat down to the last tin of sardines, the only pot of apricot jam and a tin of biscuits. The next dinner consisted of rice and salt.

With difficulty a passage was dug out to the stable and a little fodder got out to the starving horses. The dogs were another matter — their kennels were entirely buried and beyond reach. And there was no sign whatever of the sheep.

On the fifth day the cows were found and manhandled to the horse enclosure where, after a labour of four hours, some oats were got to them. By this time there was not a trace of food in the house. 'The servants remained in their beds, declining to get up, and alleging that they might as well "die warm",' wrote Lady Barker.

Thoughts turned to ripping up the floorboards in search of a cat they knew was trapped under the house, but the idea was put off — 'chiefly, I am afraid, because she was known to be both thin and aged'. Then Fate relented. Rain began to fall and the cold southerly wind was replaced by a warm nor'wester. Soon Lady Barker was able to slide on over her own boots a big pair of kangaroo-skin riding boots greased with weka oil and venture out with the others to see what could be done about the sheep.

They were met by a grim sight. Many in one mob lay frozen in the snow; others had been drowned by overflowing water as they huddled for shelter in against the raised bank of a creek. Those that survived were painstakingly dug out and led to safety. All of the sheep in a second mob were found to be dead. Lady Barker estimated that half of Broomielaw's flock and almost all of its lambs had been killed by that winding-sheet of snow.

If anything the storm was more severe in the back-country ranges, and even on the plains, where it was relatively short-lived, sheep drowned under high river banks, or came to grief against wire fences, where they were soon covered over by snow and trampled to death. News reports later put the death toll among livestock in Marlborough alone at half a million.

The weather still found ways of playing up at Erewhon. It was now autumn, and normally at this time the rains would be giving the grass a last burst of growth. This year the rains would be late, Perry had told me, and the pastures wouldn't respond. It was likely to be a harsh winter.

Events were to prove Perry right. Not long after my stay at Erewhon,

Chrissie phoned with news of the autumn muster. For the first time in eight years 'dirty' snow from the nor'west had fallen, leaving the men to push as best they could through chest-high drifts. One musterer lost his footing on a slope, falling and badly smashing himself up on rocks. With true high-country grit, he managed to get down off the hills, find a horse and ride back to the station to be airlifted out. The sheep, meanwhile, were trapped up in the summer basins. For once, a chopper was being contemplated to take the musters back in.

Now, as I write this (early in that fateful season when almost a century and a half ago Lady Barker waved goodbye to the departing dray) the South Island is bracing itself for a fresh fall of snow and some 1400 Canterbury households are entering their 11th day without power. The winter chill, which set in weeks early, has turned southern regions into a vast ice-rink, triggering record power use. The offshore storm system responsible for it all has made itself felt on the far side of the Pacific, sending wave surges to batter beachfronts from Peru to Mexico.

But all this lay well in the future as I rode back to the stables on that overcast afternoon. Colin and Perry had taken themselves off to Timaru to see a doctor about a couple of niggling injuries. Colin had aggravated a shoulder injury dating back 20 years to when a horse fell on him, and one of Perry's ankles was playing up.

CLARK SCOTT

I unsaddled Eddie and gave him a bucket of chaffed oats, watching his eager muzzle fill the container, blowing up a cloud of flakes with every breath. This year the station had sown 5 ha in oats — two years' worth of feed.

'When the horses are working hard in a team, grass seems to bloat them up and make them sweat,' Chrissie told me. 'We keep the grass down and use chaff with maybe a bit of hay.'

Chrissie had a more than passing interest in such things, having trained in homeopathy and as a nutritionist. The station animals were treated homeopathically and fed good tucker. Erewhon's hens and dogs, and even the house cat, were raised on something called Mighty Mix — a wholefood concoction of Chrissie's devising. Indeed, it was largely thanks to Mighty Mix that the Drummonds were on Erewhon in the first place.

The mix started life at Kekerengu Station near Blenheim in the winter of 1992, in the midst of a biting snowstorm. Despite regular feeding, the working dogs had become malnourished. The sheep had lost condition. Everything that walked upon the earth got to be in a very sorry state. While Colin was away snow-raking alongside Army regulars, Chrissie stirred herself to action, throwing together a natural food from whatever suitable ingredients she could get her hands on.

'Within ten days my dogs had bloomed,' she said. 'Then, thinking of the state of others round about, I started to feel guilty. There was a big problem with sheepdogs collapsing.'

> For the first time in eight years 'dirty' snow from the nor'west had fallen, leaving the men to push as best they could through chest-high drifts.

So she made bigger batches of what grateful neighbours soon christened 'Mighty Mix' — three tonnes at a time, in 60 kg sacks — and carted it around the district as relief feed. That put her on the news. Word spread, demand increased and the dog-food business grew. She got herself a factory in Blenheim, hired a sales team and began selling the product nationwide. Income from Mighty Mix eventually helped bankroll the purchase of Erewhon.

Chrissie being who she was, the story didn't end there. A new and even more impressive chapter began one night when she struck up a conversation at a social evening in Christchurch. During the course of the conversation Chrissie learned of the dire condition of villagers on Lake Victoria in Kenya. The Rusinga Islanders had been finding it increasingly hard to live off the lake's elusive fish stock, and to make things worse, the impoverished community was being ravaged by HIV/Aids, which affected one in three of the island's population.

If natural wholefoods could energise high-country sheepdogs, surely they could do something for the poorly fed islanders. Before long, Chrissie had created New Zealand's Nourish Seafood Delight — freeze-dried food that did not need enzymes to break down and so could be absorbed directly into the bloodstream. The month I was at Erewhon, 4000 lunch packs of Seafood Delight had been despatched to Rusinga Island. A month earlier, the islanders had sat down to Beef Tea, and the following month Chrissie planned to introduce them to Lamb and Mint. Money from the dog-food business also helped underwrite a homeopathic dispensary and fund a programme of malaria and chickenpox vaccinations for the 20,000 islanders.

> 'A lady's influence out here appears to be very great, and capable of infinite expansion . . . and her footsteps on a new soil such as this should be marked by a trail of light.'

'A lady's influence out here appears to be very great, and capable of infinite expansion . . . and her footsteps on a new soil such as this should be marked by a trail of light,' Lady Barker had written 140 years ago. She had in mind everyday station life, but today Chrissie — and women like her — continue to test that notion of infinite expansion in an age of global awareness.

Lady Barker went on to describe her daily routine at Broomielaw. She would be out early with a tin of scraps to feed the hens and ducks, then back and dressed for morning prayers and a breakfast of porridge followed by mutton chops, mutton curry or perhaps broiled mutton and mushrooms, along with fresh eggs and coffee. Then came a visit to the kitchen to oversee preparation of the midday meal, which was a full-on affair involving soup, a joint, vegetables and pudding. That was followed by an hour or two writing then, later in the day, a long ride of perhaps 50 km to work off all that time spent at table, culminating in a return home at twilight.

On one such ride, after crossing a low range of hills, Lady Barker chanced on a 'nest of cockatoos' — a settlement of small freeholders. They seemed to be doing well enough but she was dismayed at the lack of education for their children — the nearest school or church was 50 km or more away. So Lady Barker did what others before and since have done: she set about remedying the situation by fundraising, and soon had the pleasure of seeing a local schoolroom built, in which church services could also be held.

Pondering how life on the land repeats itself in the essence of its unfolding from one decade — from one century — to the next, I got talking to the real boss of Erewhon, its cook, Annette Girvan. Annette and her husband Brian, Erewhon's handyman, once cooked for the shearers at Mt Algidus station. 'I've always wanted to work there, having lived in a street named after it in Christchurch,' said Annette. 'After being there a while you could understand why Mona Anderson wrote those books.'

The station was later put on the market for $18 million. 'I believe the owner got $14 million,' said Brian. 'He was there 12 years and made a million dollars a year in appreciation.' A buyer in Scotland bought Mt Algidus and put a manager in. 'What is produced here wouldn't even pay the interest on its value,' he said.

Brian's uncle had worked at Erewhon as a guide at a time when the property incorporated the adjacent Mt Potts station and falling profitability had forced a change of direction. Brian handed me a printed foolscap sheet headed 'Erewhon Park Limited' and dated June 1975. On one side was a map of the central South Island demonstrating how accessible the Park was — just 168 km from Christchurch via the Rakaia Gorge, and 136 km from Timaru. Reasons for making the journey, listed on the reverse, included hunting safaris for red deer, thar, chamois, rabbit, hare, possum, Canada geese and practically anything else that moved (all taxidermy to the highest overseas standards); trout and salmon fishing, horse treks, scenic flights from a licensed runway, educational tours and skiing at Powder Snow Valley. Accommodation offered at the Park, which was headquartered off the access road 9 km short of Erewhon's homestead, included camping sites, a bunkhouse and rooms at the lodge.

An Intentions Book by the front door of Erewhon's homestead marks the property's swing back to high-country sheep farming. A notice on its cover welcomes hunters and climbers who have had the courtesy to ring and ask for access, but carries a stern warning for those who haven't, and are therefore 'trespassing and in danger of stuffing up access for everyone else'.

Brian's career itself bears testimony to the changing nature of sheep farming. In the halcyon days of the 1960s, when around 75 million sheep chewed New Zealand grass, he joined the ranks of the wool classers — a specialised breed of men who sort the clip into any one of 360 wool types, scrutinising it for subtle differences in strength, size and purity.

'I loved it,' said Brian, 'thought I was made. They used to reckon there was no difference between wool classing and being a doctor — it was so highly skilled.'

I have before me as I write a 1961 printing of *Classing the Clip*, first published in 1928. A handbook for practitioners, it gives the uninitiated a glimpse of the bewildering complexity of the fibre on the sheep's back. Besides the obvious qualities of length and fineness, classers are instructed to assess tensile strength, elasticity, felting capacity — 'the power to felt, or entangle, so as to form a compact tissue' — softness, pliability, density, character — 'generally recognised by the development of the crimp' — and colour.

Colour itself ranges from full-lustre (a rich, glossy metallic sheen typical of Leicester and Lincoln breeds) and first demi-lustre, to half-lustre (common with half-bred Lincoln-merino) and second demi-lustre (an almost white wool associated particularly with Romney Marsh sheep). The wool can also become yolk stained, discoloured by urine or permanently greyed by the excreta of tick parasites. Then there is the advisability of keeping separate the wool from the different flocks — ewes, wethers, hoggets, rams and lambs.

Environment also played its part: high-country merinos produce a sandy wool where it gathers wind-blown dust, high-rainfall regions result in yellowish wool, and rich pastures are marked by a greater amount of 'yolk', or natural grease in the fibre.

All of this rarefied knowledge was brought to bear in the woolshed where, under the same roof as the sweating pens, catching pens, chutes and shearing stands, the classer laboured to impose some order on what the sheep had grown unsupervised out on the hillsides.

Having organised the table-hands and then, once the shearers got busy, allowed a good number of fleeces to accumulate on the sorting table, the classer 'fixed the standards' by deciding on the number of grades the clip would be sorted into. Shearers being past masters at picking the easiest sheep to shear — which happen to be those with the best wool — the grades often had to be rejigged as the less impressive fleeces came through.

Thanks to a never-ending squeeze on farm profits and the move toward sheep bred for meat, the handbook's descriptions of finely honed sorting tend to be in the past tense. 'When I first started, the classing was so precise,' said Brian. 'Then more and more wool was grouped together to cut costs, or not classified at all. The result is a decline in wool quality and poorer quality garments.'

A few years ago Pyne Gould Guinness (now part of PGG Wrightson — the country's only nationwide wool broker) had five wool stores and 18 classers in Christchurch. Then the company relocated its trade to Dunedin, leaving just one Christchurch wool store. The number of sheep displaced by grapes north of Canterbury had made the southern city more central for warehousing. The upshot was that after 30 years in the game Brian was made redundant.

In chaos theory it is known as the butterfly effect — a small event in one part of a system starts a chain of actions that results in a large and unexpected consequence somewhere else. Plant a vine in Marlborough and a lifelong wool classer finds himself on Erewhon Station.

The butterfly effect is especially prevalent in the rural sector. Ruck Keene lets a handful of rabbits go in Nelson in the late 1850s and 20 years later he is a ruined man. A year or two after the rabbits have been released Lord Petre of Essex and Prince Albert, the Prince Consort, ship out a few deer to Nelson and a century and a half later the animals are being shot at one end of Erewhon and mollycoddled at the other.

Colin told me that when Arthur Urquhart bought the station in the 1930s it was overrun with deer. Urquhart set about culling them in a determined sort of way and for the first three years he was said to have earned more from deer skins than from the woolshed. By dramatically reducing the number of deer he released a phenomenal amount of grazing potential out in the hill country basins.

'Over the past few years a lot of the higher grazing land has been getting choked up with mountain scrub,' said Colin. 'It's amazing the pressure those big herds of deer had on everything.'

The Drummonds are understandably keen to allow hunters into Erewhon — especially those targeting thar, which though often thought of as deer are actually goats. The first Himalayan thar to arrive in New Zealand were gifted by the Duke of Bedford in 1904 and released in the Mt Cook district, not far from Erewhon, as were the first introduced chamois three years later.

Colin claimed that thar were the most destructive animals he had ever encountered. The rise of summer hunting was good, he said, because it kept them moving and discouraged their natural inclination to congregate in large mobs.

Rakaia red deer are another proposition entirely. Erewhon has 450 of them — all acquired from wild capture — penned on the flats and grassed hillsides at the front of the station. The reds are renowned for their bodyweight and for the quality of their velvet, but only a few herds remain. The rest have introduced European genetics. A lot of European deer are larger, but they are high-performance animals with big appetites. 'Because the Rakaia reds have been around here for so long, they have adapted to a simple low-cost structure,' Colin told me. 'There is a big difference between high-performing and low-performing deer.'

It works like this: on lowland estates with expensive pastures and high maintenance costs the income possible from velvet and venison often makes the profit equation problematic. By contrast, Erewhon's winters kept sheep and cattle returns down to $40 to $45 per stock unit. So deer were still 'right up there' in terms of viability.

The next morning I got a chance to see the reds up close when I was feeding out in one of the paddocks — anyone visiting Erewhon is likely, before long, to be directed toward some pressing piece of work (on the way back from refilling the hopper I had caught sight of Colin's father on a tractor, out discing a field).

The deer trotted eagerly toward the truck, forming a long, sinuous line and bending their delicate heads to the trail of grain as it fell to earth. Nearby, the tendrils of the Rangitata snaked among the stones of the riverbed, which at this point was around 4 km wide. The distant skyline was crowded with the peaks and ridges of the Two Thumbs Range, their shape mimicked by a cover of cloud.

Downstream, a 100 m high hummock rose from the Rangitata's stone plain. It was Mt Sunday, a moody piece of ground that had tasted brief fame as the ancient walled city of Edoras, a stronghold of Men, when New Zealand film crews laboured to put flesh on J.R.R. Tolkien's *The Lord of the Rings*.

As I looked, more deer entered the pen from the low scrub, the ever-present dogs at their heels. Colin rode up behind the mob alongside Perry and the station hand. All were on Clydies.

Of course the gentlemen were very fussy about their equipments, and hung themselves all over with cartridges and bags of bullets and powder-flasks; then they had to take care that their tobacco-pouches and match-boxes were filled; and lastly, each carried a little flask of brandy or sherry, in case of being lost and having to camp out.

So wrote Lady Barker of a wild cattle hunt. Colin and the others could have ridden off her handwritten page and straight into the image-making machinery of the Southern Man. The inestimable woman herself preferred to carry a flask of cold tea and a stout walking-stick.

Catching my eye as he came up, Colin smiled warmly under a battered broad-rimmed hat. Perry leaned across, and with a straight face delivered a bawdy observation that had us all laughing like drains. At his back, the dusty road eased over a rise and dropped from view, on its unhurried way to the distant glass-towered formalities of Christchurch.

At the other end of the Rangitata River, a few kilometres from where Richard Pearse, the Birdman of Upper Waitohi, fitfully broke the shackles of gravity at the dawn of the age of flight, crop farmer Rodger Slater was on all fours crumbling dry earth between hardened fingers.

Hectares of barren soil stretched out around him, covering the wrinkled, inactive pea seeds. Elsewhere, canola, linseed and buckwheat enlivened the land with a palette of colours. Rodger stood and brushed the dirt from his hands. There was unlikely to be any establishment in that field. 'This is abnormally dry,' he said. 'It is getting quite serious. If we get significant weather changes, what will it be like in twenty years? I must admit, that thought niggles away in the back of my mind.' He nodded to the inland hills, a few kilometres away. 'Up in that country they are starting to feed out already.'

'They' were sheep farmers. Once, they would have had a paddock of grain and a small harvester along with sheep and a few cows. Now they had specialised, and — like all specialists in nature — they had become vulnerable. As they nervously eyed their shrinking stock of winter feed, Rodger's phone had started ringing.

This year the rising cost of imported grain due to a weakening dollar had even sharpened the interest of local millers. Unfortunately, Rodger's grain was all under contract, so he couldn't capitalise by lifting the price. 'We really should have had more free barley,' he said. Earlier we had inspected a field of tall fescue, a grass-seed crop grown largely for the North American market. Rodger had rarely seen it so stressed in autumn.

An orientation drive around the 425 ha home block, which lies on undulating land between the Orari and Opihi rivers southwest of Geraldine, had given me my first intimation that cropping was a far from straightforward business. Replace animals with fields of biddable plants, I had thought, and most of the difficulties connected with working the land would surely disappear. No worrying about lambing percentages or pink eye, facial eczema or bleak autumn musters. No falling wool prices or wrangles with the Department of Conservation over management of high-country peaks and ridges. No crashing through bush after wild cattle. No question mark over dog chips, and no staggering from bed in the small hours to shift animals off a floodplain.

What could be easier than sowing a crop, keeping the birds off and watching income accumulate with every revolution of the sun? Plenty, as it happened.

For one thing, plants could be as pernickety as any animal. I learned this from Rodger's son Guy, who had all but assumed the mantle of farm manager. Guy drew my attention to something called heat necrosis, which affected potatoes when overnight temperatures were too high. Normally, during the day the leafy tops draw moisture out of the spud, and at night the reverse happens, the moisture being taken down into the potato tuber. If the night is too warm, however, the tops continue to draw moisture away, resulting in brown flecks in the flesh where cells have died.

There is not a lot a farmer can do about that. A few nights without needing the duvet and Guy had seen 80 tonnes of crop turn into cattle feed. In financial terms, a warm kiss or two had transformed a handy $250/tonne crop into a $10/tonne disaster. 'The contract with Talley's was for 900 tonnes, so we lost almost 10 per cent of it,' said Guy. 'Luckily we have enough good spuds to cover it.'

All up, 109 ha of them. The Nadines were badly hit, but not the Dutch Agrias. Guy also had two older varieties in the ground, Russet Burbank and Russet Ranger, contracted for French-fry production. The Russet Burbank had been around for 130 years, he said, and had never been improved on as a long-keeping chipping variety. In his view the Canterbury Plains, with its free-draining soils, good day/night temperatures and access to water, was one of the world's leading areas for growing Russet Burbank.

'Washington State is the only area that is better. Their growing season is slightly longer — 170 to 180 days compared with our 160 to 170 days. We can grow pretty cheap potatoes.' Guy had lifted 694 tonnes of Russet Ranger from 10 ha — the year's best yield.

The downside of a potato crop was its thirst for water and an obligatory five-year rotation cycle to avoid soil exhaustion, minimise disease, and combat what were charmingly known as 'volunteer' potatoes — small, overlooked tubers that would regrow next season.

As Guy talked, I thought back to a book I had seen weeks earlier in a city second-hand bookshop that bore what I had then imagined to be a whimsical title: *The History and Social Influence of the Potato*. The longer I spent in Guy Slater's company, though, the more insightful and to the point the book seemed. Its author, Redcliffe Salaman, had developed an abiding interest in genetics and chose the potato as his medium of research. In 1908 he was rewarded by the discovery of resistance to potato blight, and in the years leading up to the Second World War he headed the Potato Virus Research Unit at Cambridge in England. He was also said by one biographer to have done valuable work as chair of the outrageously titled Potato Synonym Committee. All of which suggested that a good career could be had from the potato.

The town of Ashburton must have thought so. Some 80 per cent of the Slaters' potato crop — mostly grown on sharefarming blocks in mid-Canterbury — along with harvests from who knows how many other plains farms, was fried and frozen there for export to Australia and Asia.

I soon came to realise that planning was at the heart of crop farming. The idea was to spread crops in such a way that early barley and tall fescue were harvested together in January, followed by wheat and later spring crops. 'It is quite a robust, bullet-proof system really,' said Guy.

Complicating matters is the need to rotate crops grown on any given piece of ground to minimise disease and restore nutrients, which turns the planning exercise into something like three-dimensional chess. 'On this country, if we followed wheat with another wheat crop, the yield would drop from eight or nine tonnes a hectare down to six or less,' he said. The Slaters therefore followed wheat and barley with a break crop such as borage, peas or canola, then with tall fescue. Home gardeners did the same when they took the old advice to alternate tomatoes, for example, with legumes.

Commercial crops of canola present special difficulties. There needs to be a 2 km radius of isolation from other yellow-flowering brassicas, such as kale or turnips, to stop bees cross-pollinating the crop. To help minimise this risk, growers lodge their intent on a local crop register. Self-seeding is also a problem. It takes seven years to rid a paddock of volunteer plants, which means that, given a 50 ha crop, a lot of real estate is needed to make a commitment to canola. DNA testing is also carried out to satisfy stringent buyer criteria. There is a lot of Roundup-ready canola in the world, and Slater Farms is GE-free.

From the cab of his ute, Rodger pointed to a field of buckwheat, a crop that he was new to. 'I'm not sure whether we've got a good harvest or not. Potentially, it is two tonnes a hectare.' He had been told it was to be wind-rowed — cut into rows to be dried in the open.

The buckwheat, a short growing-season plant, was put in after a harvest of silage for a local farmer. Not far off was 12 ha of linseed, which was doubling as a cover crop to establish next year's tall fescue. The linseed was a way of generating income from this part of the farm while the slow process of cropping the grass unfolded. Tall fescue differs from ryegrass in that it sets seed only off autumn tillers. To encourage reproductive autumn tillers and generate silage, Rodger 'got it revving' with water.

'This is where irrigation is particularly valuable,' he said. Concerns about climate change had prompted him to put a moisture probe in the ground a metre or so beneath the surface. 'It's been a good guide. These crops came under pressure much sooner than we could tell just by looking at them.'

We stopped at the farm's mechanical heart — a massive 350 m long lateral irrigator which ended in a water gun that extended the range by a further 20 m. The gantry was mounted on rubber tyres and driven by a computer-controlled motor. The sun beat down on steel and leaf, rutted track and crusted field. Rodger walked along to a hydrant and unclipped the irrigator hose, then dragged it with a tractor down to the next hydrant. When it was reconnected, he again set the machine in motion, on its mist-enveloped 2.5 km run down the paddock. At a single pass, the irrigator delivered 30 mm of water — more than four days' worth, given evapotranspiration of 7 mm a day.

'We had to cut down some of the trees my father had planted,' Rodger said, gazing down the cleared corridor through which the irrigator was now crawling with ponderous deliberation. 'It's a shame, but our consultant told us to ignore existing trees and fences and look cold-bloodedly at the shape of the property.'

The loss of the trees had been less frustrating than the massive power pylons that reared over the land. Rodger was forced to disrupt his watering each time he sent the irrigator their way to guide it around the pylons in a series of slow-motion 180-degree turns — all up, seven hours of down time to execute a 360-degree turn. 'They went in forty years ago,' said Rodger. 'We'd never agree to them today.'

Nevertheless, despite its cumbersome pathways, the new irrigator has drought-proofed the farm and introduced a measure of flexibility. 'We spent a lot of time putting kilometres of clay tiles down to drain this country so it seems ironic that we are now irrigating.'

The water was desperately needed, though — especially by local dairy farmers — and with no groundwater under the heavy clays, they had been obliged to look elsewhere for a useable source. They found it in the hills, building a dam in conjunction with the local energy company and guiding water through canals to the properties in the scheme.

Once again, it seemed, the economics of milk were instrumental in transforming the land. A few years ago this had been a region of crop and sheep farms, but the higher incomes from dairy had opened gates to the cows.

'Dairy farmers are efficient and they haven't sold out at the farm gate like sheep-farmers and croppers,' said Rodger. 'Their cooperative structure has let them keep control of production right through to wholesale.'

Guy had told me that historically, the gross from the property was about $2000/ha. Dairy farmers were grossing three times that — and doing it across the entire farm. 'When hail comes from the sky, the cows get sore backs,' he said. 'We get wiped out.'

Same old story. To survive, Slater Farms got bigger. Rodger had bought out three farmers over the years and now cropped twice the area his father had. Hard work had not blunted his zest for life though. After glancing at his watch he suggested, with a hint of playfulness, that now might be a good time to take a look at his latest project, Periwinkle Lodge.

The lodge, it turned out, sat next to Lake Jenny, named for Rodger's wife. The 'lake' was a picturesque buffer pond for the farm's irrigation which was planted round about with scarlet oaks. The loftily named lodge was a caravan, bought on TradeMe and parked where it could soak up the view.

The canal that fed the pond cut across a corner of the farm, and between it and the road Rodger had sculpted a modest three-hole golf course. Golf, he admitted, was his passion. He had played at the Denfield club for 30 years now and was getting what he called 'an old man's swing'. Jenny was already in the members-only lodge with friends when we arrived, the laughter and animated talk giving the place a warm vitality.

Rodger told me something of how the course had been built and why — well, you couldn't very well graze animals on such a small, difficult, uneconomic piece of ground, could you? I also began to see why the tall fescue so occupied his thoughts. Much of what he grew was destined for the United States, where prestigious golf courses, such as Augusta, Georgia, resowed with temperate grasses in winter. Rodger's three-hole course was sown with the same stuff —

'Dairy farmers are efficient and they haven't sold out at the farm gate like sheep-farmers and croppers.'

possibly even from the same crop — that Tiger Woods strode onto with his caddy an ocean away.

In case I began to get the idea that this was some sort of Arcadia, Jenny mentioned the neighbours. When she first arrived there had been seven. Now there were 14. 'People don't sell farms,' she said, 'they subdivide into lifestyle blocks.'

Rodger feared that the ease of subdivision would eventually compromise his ability to farm. 'You just have to look at Europe, where they have total fire bans and strict limits on noise and work hours. Who would want to live out here and hear harvesters until midnight, or scare guns when we have a bird problem?' In Europe, where they were required to mulch, Rodger said crop farmers were afflicted with a huge build-up of slugs and grass weeds. 'A good fire can solve a lot of problems.'

Next day, the first order of business was to torch a field. The upside of a drought is that it looks kindly on harvesting and on burning. Under a dazzling sun, Rodger set scattered piles of stubble alight, deftly flicking the long nozzle of his improvised flame-thrower this way and that to touch the dry stalks with a drizzle of flame. Tongues of fire leapt and crackled, sending up billows of smoke. It was hot work.

It also had the acrid tang of tradition about it. 'The effect is beautiful, especially as it grows dusk and the fires are racing up the hills all around us,' wrote Lady Barker of that early pastoralist practice to encourage new growth known as 'burning the run'. Lady Barker was no stay-at-home when the burning season rolled around. Armed with a good supply of matches, and having taken care not to have on any inflammable petticoats, she would help her husband set the tussock ablaze. Often, with one or two colluding friends, she would slip out of an afternoon to do a little more on her own account, occasionally getting singed eyelashes for her troubles.

If the weather had been dry for a time and the wind was high, they might attempt to burn a big flax swamp, which crackled 'splendidly' once it had got going, but on the whole Lady Barker preferred laying waste to hillsides — 'you get a more satisfactory blaze with less trouble; but I sigh over these degenerate days when the grass is kept short, and a third part of a run is burned regularly every spring, and long for the good old times of a dozen years ago, when the tussocks were six feet high.'

I was surprised by a mouse that darted from a patch of burning stubble at my feet and made for the nearest sheltering crop. Moments later, the big Fendt tractor that was busy dragging a set of discs across a nearby field disturbed a hare which bounded through the dust to join the mouse.

The tractor itself was symptomatic of the changing face of cropping. The Slaters didn't have a machine with enough grunt for the job, so the sophisticated piece of German engineering that was playing havoc with the wildlife had been contracted, complete with driver, from a neighbour. With tractors costing upward of $1000 for every horsepower and a new harvester setting a farm back $400,000 or more, it is not surprising that contractors, with the specialist equipment they bring to bear, have become popular — they save farmers the expense of having one of everything in the shed.

'Most people could get a business up with half a million dollars, but you'd barely get out of the blocks in agriculture,' Rodger had told me. 'You can have a lot of assets, but it is hard to crystallise the investment into income.'

Nevertheless, Slater Farms had accumulated several harvesters and tractors as well as assorted bits and pieces fitted up with spray units and other such things demanded by industrial-scale gardening.

The Farms also had a secret weapon from Yorkshire by the name of Charles. Mechanically gifted, he had good cropping credentials and was fresh from eight weeks of driving harvesters in Western Australia. The weather there was so stable, he told me, that contractors worked back-to-back 12-hour shifts out in the field. Rodger was plotting ways of luring him back next summer — maybe he would even throw in an airfare from the Old Country.

I spent time in the harvester cab with Charles as attempts were made to bring in a field of white radish seeds for a Japanese client. It was a highly technical operation. Tune the harvester too highly and the seeds were smashed about. Ease off overmuch and too many were left on the ground.

A convocation gathered around the trail of mulched chaff left by the machine. Guy sifted the leavings. Rodger sifted them. Jon Hanrahan, a field agronomist from South Pacific Seeds, did the same. Adjustments were made and another 40 m swathe was cut. Eventually, Charles got results everyone was happy with and he set about harvesting with a will.

South Pacific Seeds is what Jon called a 'seed multiplier'. It deals at the high end of the hybrid seed market for demanding clients in places like Japan, Korea and Holland. The seeds have to be viable. Too much bruising can damage the embryos and even if they germinate, they may have no vigour.

'There is not much in the way of margins and no room for mediocrity,' Jon said as he climbed up to the hopper to check the moisture content of the seeds. After fiddling about with his gauge he gave his verdict. 'Six point two per cent.'

'So no air drying, just store,' said Rodger. 'I like it.'

That night I drove with Rodger out to Peel Forest Estate, a sheep station dating back to the 1850s that lay beyond Geraldine in the direction of Mesopotamia and Erewhon. Its present owner, a Yorkshireman by the name of Graham Carr, had turned Peel Forest Estate into a deer stud of some note.

We swept up a drive lined with 150-year-old Douglas fir the like of which, I was told, existed nowhere else outside the continental United States. The other regulars who made up the weekly tennis foursome were already getting kitted out and twirling their rackets in a high-spirited manner.

The floodlit court was an impressive affair. It was fenced and screened from the homestead by a high, clipped holly hedge, and sported at one end a delightful wooden pavilion. Graham started up a vacuum cart with difficulty and majestically circled the court a few times picking up stray leaves. Then all friendship was set aside and the game began.

I eased back into an old wicker chair with a can of Boddington's and watched the farmers fight it out with grunts and calls of encouragement. Somewhere a stag roared into the night. The ball went back and forth with a satisfying thock. Now and then it landed in the pavilion and, setting down my ale, I retrieved it and heaved it back toward the net.

Over dinner, talk would turn to the uncommonly dry weather, the dire outlook for sheep and deer, and the farmer who, only half-jokingly, wondered whether he would be forced to begin eating his way through his flock.

Such thoughts were best left until the morning. For now, under the high lights, among friends and with a game to attend to, this was the life.

⬆ Erewhon Station sits at the headwaters of the Rangitata River, beneath glaciers of the Southern Alps.

⬅ Horse paraphernalia fills stables which are at the heart of the traditionally run station.

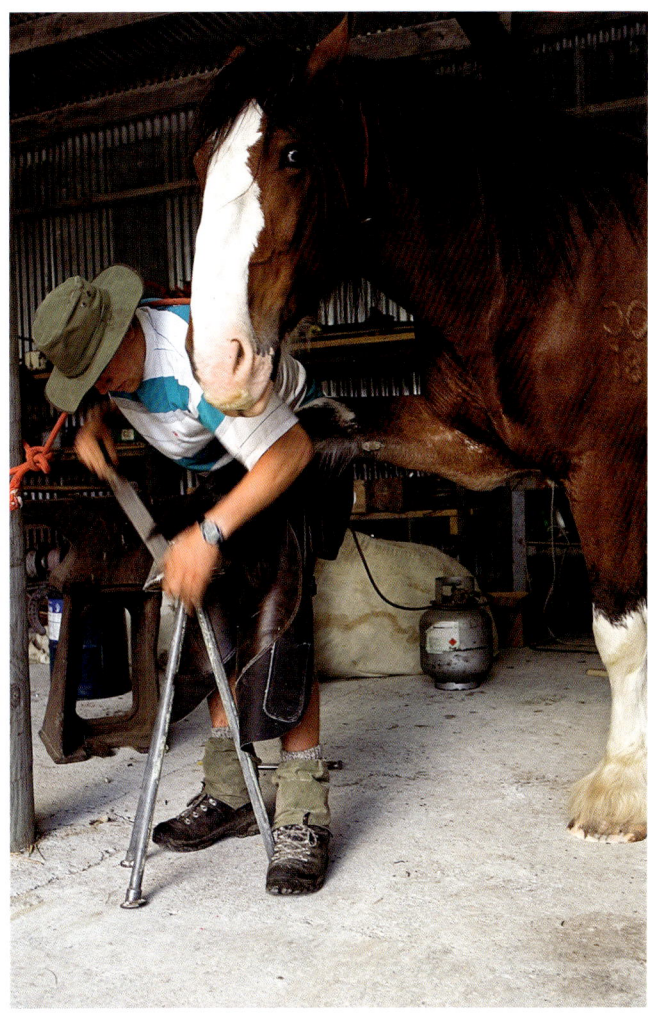 

◂ Station owner Colin Drummond strains to reshoe one of the Clydesdale workhorses. Farmhand Marthinus Vermenten is learning the once-common task, now seldom seen on the big runs.

◂ The imposing but good-natured Clydesdales find all manner of work at Erewhon, from mustering and pulling wagons to packing in supplies to outlying huts.

◂ Scree slopes and towering ridgelines introduce an element of mountaineering to stockwork at Erewhon.

▾ The stony beds of the Clyde and Lawrence Rivers dictate the use of horses in the back country. The impracticality of putting up fences here makes old-style boundary keeping a regular chore.

193   *the* Life

Kennels cut from fuel drums exemplify the inventiveness fostered by isolated rural life.

Erewhon's hermitage hut on the Lawrence River offers welcome shelter in a harsh, unforgiving environment.

199 the Life

↑ Christine Drummond tends a hearty evening meal at the hermitage's fireplace. She and husband Colin have laboriously restored the huts on their property.

→ Christine fits a protective bootie to one of the station's sheepdogs. She now makes and sells them to other farmers working in similar extreme conditions.

↓ A rider on a sure-footed Clydesdale herds cattle along a highway of stones to fertile pastures upriver.

⬆ A truck with winter feed crosses Erewhon's front pastures. The sheltered flats provide a winter refuge for the station's merino sheep and cattle.

⬇ The farmhand's kitchen offers a few home comforts for Marthinus Vermenten, an agriculture student at Lincoln University.

⬅ Erewhon has one of the country's few herds of genetically pure Rakaia red deer. Famed for their bodyweight and the quality of their velvet, the animals thrive on the station which, like many farms, is diversifying in order to remain viable.

⬅ Working the land in a time-tested way, Christine and Colin Drummond feel a kinship with earlier generations of high country farmers.

211 *the* Life

215 *the* Life

Light momentarily enamels a peak above Lambert Glacier before another front approaches Erewhon.

Experienced hands test a barley crop nearing harvest on Slater Farms outside Geraldine in Canterbury.

219 *the* Life

▲ The Slater family home overlooks a field of ripening linseed. It was built on high ground away from aggravating dust by Rodger's asthmatic father.

◀ Rodger Slater fine-tunes a combine harvester to work a field of borage. Crop farming is dependent on increasingly sophisticated and costly machinery.

⬆ While barley is a mainstay of cropfarming, a surprising number of seeds are harvested at Slater Farms, including tall fescue for golf courses, borage for oil, and white radish for the Japanese market.

➡ Jenny Slater approaches country life with flair, lending a sense of occasion even to a harvest tea-break.

⬇ Flowering buckwheat carpets a corner of the 425 ha home block. A succession of crops colour the landscape like an Impressionist canvas.

Farm 230

▸ Rodger Slater's son Guy disturbs the farm's tranquility with a brief firestorm. Post-harvest burnoffs enrich the soil and prevent the buildup of pests and unwanted weeds.

▸ A 350m-long computer-controlled irrigator creeps forward on its ponderous 2.5km run. Water is the farm's lifeblood and Rodger has invested heavily in a joint scheme to 'drought proof' his property.

➡ Fresh from harvesting in Western Australia, Yorkshire-born Charles Moverley brings valuable engineering skills to the farm. With a new combine harvester costing upwards of $400,000, regular maintenance of the specialist equipment is vital.

- Crops are analysed for moisture content and maturity to determine the optimal time for harvest. Once the crucial decision has been made, staff work long hours to take advantage of what might be a fleeting window of opportunity.

- Weather dependent and with fine profit margins, crop farmers rely on rigorous planning and years of experience to overcome the vagaries of nature and fickle global markets.

235 *the* Life

237 *the* Life

F*the* ut

A few kilometres north of Huntly, on a narrow strip of land bordered to the east by Lake Waikare and to the west by the Waikato River, State Highway 1 cuts irreverently through the old battlefield of Rangiriri. It was here, more than 140 years ago, that beleaguered Maori took a defiant stand against 1500 advancing British troops. Their defeat in what was the decisive clash of the Waikato War sounded the death knell for Maori resistance in the rich Waikato heartland and opened the way for its occupation by settlers.

Today the fighting is a thing of distant memory, and this land, with its lush, rolling pastures and benign climate, has long since been transformed into one of the country's prime dairying regions.

Just up the road from Rangiriri and its cemetery of war dead, I took the Te Kauwhata turn-off around the lip of the lake and made for Steve and Julie Nelson's patch at Waerenga. I was keen to see how things had changed since my childhood stays on farms where my brother was sharemilking. Those far-off days survive as fragmented images — sitting astride the farm nag to follow cows plodding in from the fields; hosing down the concrete floor of the milking shed to the crooning of a big valve radio, incredulously watching a bull at work.

I found Steve and Julie at the table with two of their sons, picking over the remains of a country-sized lunch. Joseph had left school and was going into the building trade. Daniel, in his mid-twenties, had turned his back on surveying a year earlier and was home again as a fifth-generation farmer. Steve himself had set that particular work pattern.

'When I left school I swore I would never milk another cow,' he told me. For seven years he was as good as his word, fronting up each day to a draughting board in town. Office life didn't come easily though, and eventually he gave it away and returned to farming. It was a decision he had never regretted. In his former incarnation he did one thing — put buildings on paper. Here on the farm there was so much variety, so many things to get a head around, that a person felt continually renewed. It was a long list: animal health, soils, diseases, machinery, staff management, bookwork, technology, changing farming practice . . .

For the past 17 years Steve and Julie have lived in Waerenga, getting through that list by following a routine whose outline I recognised. Up at 5.30 for the morning milking, which finished some three hours later with the shed wash-down and the feeding out of maize or palm kernel extract into the feed bins or, during summer, grass silage in the paddocks. Then back home for breakfast, followed by any number of farm chores — from silage-making and the spreading of fertiliser to drenching and equipment maintenance. All of which took a diligent cockie through to 3 o'clock and the start of evening milking.

'It's the afternoon milking that gets you,' Steve said. For years he had trod the path to the shed morning and night. Then, 12 months ago, in an effort to back out of the relentless routine, he had promoted himself to relief milking — working weekends, holidays and through calving.

When the children were young, the 3 o'clock deadline was a tyranny, said Julie. 'Whether it was Christmas Day or the kids away at rugby, we still had to be back for milking.' Now, if life gets in the way of farm commitments, there are others around to shoulder the load, including farm manager Dave Mellish.

For many cockies, especially those on the foothills of farm ownership, there is often no relief, and some have been known to go 10 months without a day out of the cowshed. But if the work was hard, the rewards were significant. Dairy farmers enjoyed a good lifestyle, were not accountable to others, and had the chance to increase their capital. Steve put a caveat on that sentiment. Twenty years ago it was a way of life, he said. Now, first up, it was a business.

That was a view my brother would no doubt share, if he was still out in the field. New Zealand currently has around 14,700 dairy farmers and 2.8 million cows. Two farms in five have some form of sharemilking arrangement — that old practice derived from Scottish sharefarming and American sharecropping, whereby an outsider contributes labour, know-how and a little capital in return for a slice of farm income.

As dairying has become more complex and highly tuned — more business-oriented, Steve would say — sharemilkers have found it increasingly difficult to cross the divide from labouring to farm ownership. A major obstacle is the spiralling cost of land. In 1972 a farmer needed the equity represented by 256 cows to finance the purchase and stocking of an average-sized farm, around 80 ha. By 1997 that figure had soared to 980 cows, and today land value is at a record level relative to the income that can be got from it.

Then there is the double-edged sword of the co-operative system. While the industrial muscle of New Zealand's largest dairy co-operative, Fonterra, gives farmers the strong base needed to compete internationally (and 95 per cent of dairy production is exported) its dominance means that there is little room to manoeuvre for its three smaller domestic rivals, and therefore little local competition. Moreover, Fonterra requires all of its dairy farms to own shares in the company, based on the volume of milk supplied. For the average farm this locks up around $600,000.

Steve can see the logic of the co-operative structure, however. Struggling farmers in Britain and many European countries are seeing their income squeezed by the all-powerful supermarkets, he says, because they do not have the bargaining power that comes from being shareholders of milk companies.

A new piece of sharebroking sleight of hand, called the Dairy Equity Fund, has the potential to free up Fonterra money and allow farmers to put it to good use closer to home. The fund would pay farmers the cash value of their shares, but they would retain ownership and voting rights. In turn, farmers would be obliged to pass on to investors the full value of shares, including any capital gains, when they were sold. It is likely that the closed shop of the co-operatives will be further prised open as competition increases and new financial alliances are created.

In dairying, nothing stands still, and even Fonterra's grip on the industry is likely to loosen. While I was in Waerenga, Fonterra declared a record number of applications from prospective new co-op suppliers and a record production season — a staggering 1.1215 billion kg of milk solids. The month of my visit, dairy accounted for 19.6 per cent of the country's export earnings and totalled $5.4 billion for the year.

Nevertheless, Fonterra's exclusive access to some export markets, courtesy of the Dairy Restructuring Act of 2001, ends in 2007. If the telecommunications industry is any guide, when the props of privilege are removed, eager and inventive alternatives tend to blossom. Early in 2006 further change was signalled when Landcorp, the country's largest corporate farmer, announced that it was redirecting almost a quarter of its milk production to a Fonterra rival, Waharoa-based Open Country Cheese, a three-year-old fast-growing exporter.

Sitting at the Nelsons' table, I reminded myself of how dairy farmers had embraced technology with the same enthusiasm as sheep farmers and, through the blind whimsy of Fate, now had the more favoured prospects. The refrigerated cargo aboard the *Dunedin* in 1882, which famously launched the sheep industry in New Zealand, also included casks of butter and cheese. The safe arrival in London of dairy products from the far side of the world — from what would become the Empire's pantry — was the slow fuse that set dairying alight. Although the pastoralists reigned for a time, refrigeration, centrifugal cream separators, milking machines, stock-breeding programmes, co-operatives (the first was at Springfield, Otago in 1871) and even the application of cobalt to remedy 'bush sickness' in the central North Island gave dairying an inexorable momentum.

One of its strengths, as Steve was not slow to point out, was the range of end-uses for its products. Aside from beef and the array of products crowding the chilled shelves of supermarkets, there is a growing demand from the pharmaceuticals industry. Even the dairy staples are continually being tweaked. The latest — a runaway success — was chocolate-flavoured cheese for markets in Asia.

'What do sheep farmers have?' Steve asked. The question was not cruelly meant, though the short answer was sobering. They had wool and meat. It would take an unusual, though by no means impossible, roll of the dice to reinvigorate the future of either.

Cows now infiltrate the precincts of sheep in the deep south and mix it with crop farmers on the plains. They are also taking on that other settler favourite, the Monterey pine, *Pinus radiata*. Originally planted along with the Monterey cypress (macrocarpa) as windbreaks and for fuel, pines readily self-seeded. By the 1950s their commercial potential had been harnessed and the State owned Tasman Pulp and Paper company was set up to process timber from the vast Kaingaroa plantation forest bordering Urewera National Park east of Taupo. Now Landcorp is transforming much of that once-profitable timber country, converting 25,000 ha of it to dairy and trucking in some 30,000 cows.

Nevertheless, the cockies are not having it all their own way. In addition to the usual vagaries of the market, they face increasing pressure from what Steve calls 'outside influences' including, in the Waikato, lifestylers and market gardeners. The tenant of a house on Steve's property travels daily to her job at an Auckland hospital, and like-minded commuters have begun settling in the district. For others, the rich Waikato soil is the magnet.

'Across the road, it's all spuds and onions,' Steve said, leading me out into the thin afternoon sun for a closer look.

> On the surface, the scene was unchanged from that of my childhood. The aroma of defecating beasts hadn't been improved on — despite much name-calling across the land and many acrimonious exchanges over gaseous warming.

On the surface, the scene was unchanged from that of my childhood. The aroma of defecating beasts hadn't been improved on — despite much name-calling across the land and many acrimonious exchanges over gaseous warming. Neither had the tendency of flies to enjoy dairying and the opportunities it offered. The grass still grew green, and the iron-clad shed still beat to the urgent pulse of the vacuum machine. Dry, cracked mud clung to ribbed tractor tyres as in the old days, and curing silage still exuded its sweet perfume. True, the Nelsons got by without dogs but, companionship aside, it was hard to see what a dog — even one that was supremely motivated and intelligent — could contribute on such a place. There wasn't even the prospect of an evening's pig-hunting.

Behind the scenes, however, plenty had changed. Most of it was to do with micro-management. Dairy farmers were in the business of turning grass into a versatile white commodity, and much science had been brought to bear on the process in recent years.

'Twelve years ago we basically fed the cows grass,' Steve said. 'If we had a tough summer, by March there was little left in the field and we were having to dry cows off. So we were getting maybe 200 or 250 milking days from a cow.'

Then farmers introduced maize silage, the diet was better managed and cows were fed optimally year-round. Summer grass is high in fibre and carbohydrates. In spring, the fast-growing pasture is wetter and higher in proteins so maize silage is fed out to boost carbohydrate intake. At calving, calcium and magnesium are added to drinking troughs or directly to the maize to overcome the deficit in trace elements that can cause a condition known as milk fever.

It all stemmed from the relentless pressure to increase milk yield. 'As with any industry, if you stand still, you end up going backwards,' said Steve.

One tool in his armoury was the plate count metre, a probe that determined the dry matter — the biomass — in a pasture. Dry matter, in turn, was a measure of the amount of feed available to the herd. A reading of 3000 kg of dry matter per hectare meant that the cows could be given 12 hours in the pasture to get sufficient food into them and still leave 1500–1600 kg/ha behind — the ideal amount for rapid grass growth.

Changing weather patterns mean that grass must be managed differently from one year to the next. A wet winter will necessitate working paddocks in a way that minimises surface damage, and some years there is clover flea or stem weevil to deal with.

When Steve moved to Waerenga from Morrinsville, he brought his Friesians with him. A mainstay of New Zealand farming, the big-uddered Friesians put more milk down than the lightly built Jerseys. They also produce bigger offspring, which means a better return on calves. Nevertheless, the farm originally ran Jerseys, which Steve acquired when he bought out his brother eight years ago. They make up a third of the 650-strong herd and they too have their strengths, not least a higher percentage of milk solids and an ease of handling when it comes to drenching and milking.

A new animal — the 'Kiwi cow' — has also begun to make an appearance. A Friesian-Jersey cross, the black or black/brown hybrid is symptomatic of the restless search for high-performance stock that has characterised farming since colonial days.

It is a fact of genetics that the first hybrid of a cross will be high-revving, with improved production characteristics, but for Steve it does raise a question: 'Where do you go with that calf?' The Kiwi plays havoc, too, with farm aesthetics. 'With a cross, you end up with a herd of all sorts — tall, short, different colours.'

Many dairy herds in the United States and elsewhere are fed low-cost grain indoors and only calve every second year. New Zealand herds don't have it so good. Cows here are got into calf each year before the flush of spring grass. Two months prior to calving they are dried off to allow their bodies to adjust. This has the benefit of reducing the demand for expensive summer feed.

For the past four years Steve's cows have calved in mid-June, but he is in the process of introducing split calving, with around 250 cows calving in early April and the rest toward the end of July. In winter, grass grows better at Waerenga than further south where the temperatures are lower — Steve even noticed the difference between Waerenga and his old farm at Morrinsville, 50 km away.

'We use winter grass to put milk in the vat. Winter milk attracts a marginal premium,' he said. 'We try to get 300 days of milk out of our cows, and we do that with the best grass we can grow, which is still the cheapest feed, and use supplements such as maize when there is insufficient grass.'

Like many farmers, Steve employs a consultant who visits the farm monthly to look at production, feed levels and supplements, and to help with forward planning. For the past six or seven years he has also been part of a financial group which collectively analyses data to gauge the performance of farms in the group.

Like businesses under pressure everywhere, one response among farmers has been to grow. They have added cows — the average herd size has grown from 150 cows 10 years ago to almost double that today — and they have added farm.

Steve drove me beyond his old property boundary, past a stand of kahikatea and along a metalled lane flanked on either side by bare fields. The soft brown rock under our wheels, 270 truckloads of it, had been brought in to form new access-ways — anything sharper would have damaged the cows' feet. Buying the land had rescued it from what Steve might term 'outside influences'. The previous owner, who intended growing potatoes and onions, had used a bulldozer to clear the slate with a thoroughness that General Foch would have admired. Having purchased it, Steve fitted it once more for the noble pursuit of dairying by re-fencing and sowing ryegrass and clover.

He admitted that the new property couldn't have become available at a worse time. But then, a farmer had to look steadfastly into the future and weigh up the likely viability of a farm 10 or 15 years down the track. Adding hectares could only help matters.

Aside from the adjoining land, Steve also leased two local blocks as run-off farms and was involved in a syndicated farm at Omahuta in the Far North, halfway between Kaikohe and Kaitaia. It was difficult country up there, he said. Hotter and drier than the Waikato, and with poorer pasture and the curse of kikuyu. Land tended to be hilly, and the heavy soils meant that pastures were easily damaged by cattle in winter.

Turning back to the original farm, we passed under a line of pylons which put me in mind of the grief similar pylons had brought to Rodger Slater down in Canterbury. Three lines went through Waerenga, two dating back 60 or 70 years, and one that Steve remembered being put up when he was a lad, barely in his first gumboots. 'Our grandfathers felt the lines were for the betterment of New Zealand, but they didn't like them,' he said.

Time had not mellowed the discontent. Trees couldn't be grown under them. They were noisy. They generated magnetic fields. 'And they have upped the voltage on this line above us.'

Worse was about to come. National grid operator Transpower planned to run a new 44 kV line from Whakamaru in the Waikato to power-hungry Auckland in the near future. The controversial 186 km line would pass the Nelsons within two kilometres of the existing lines. To sugar-coat the bitter plan, and to win Electricity Commission backing, Transpower had offered to run the line at 220 kV, only boosting it to full power around 2020. This was nowhere near good enough for the opposing landowners, including the great many lifestylers whose blocks were affected. They vowed to fight the power company all the way through the lengthy Resource Management Act consent process. It was a heartening act of solidarity between cockies and newly ruralised city folk.

A similar, and equally unlikely, meeting of minds had occurred over proposed legislation to introduce the mandatory microchipping of dogs. In the first significant defeat for the minority government, a number of Green Party members of parliament voted against the proposal and forced an amendment exempting working dogs — the vast majority of them farm dogs.

Living less than a kilometre from the scene of the vicious 2003 dog attack that prompted the introduction of the bill, I could sympathise with those who wanted comprehensive microchipping. Nevertheless, having spent time with literally hundreds of farm dogs, I would have to put it on record that, as breeds — as individual animals — they are too much governed by the Protestant work ethic and the joys of country living, and in any case too removed from built-up neighbourhoods, to cause much offence.

Blanket microchipping was a bureaucratic response, and was criticised as a way of avoiding taking unpopular measures to control the few inherently dangerous breeds. It would also have been ineffectual, in the way that the lock on my front door is ineffectual, except as a deterrent to honest people.

Farmers, understandably, took a different view on microchipping. Talk from the rural quarter hinged on the cost to a bloke who had a posse of dogs, and on the legislation as yet another tax and an unwanted addition to the red tape that was growing to unwieldy proportions.

More justifiably, farmers are coming under pressure to modify their environmental practices. Animal-rights activists have long opposed docking and the inducing of cattle, and have bemoaned the lack of shelter afforded by typical pastures. Like many dairy farmers, Steve no longer induces cows or regulates their menstruation with drugs. Nor is he an enthusiast of chemical sprays, though he does employ them 'with caution'.

'Sprays help us remain commercial. If we didn't use them, we would face a lot of substandard weed-affected pastures,' he said. 'It's frustrating to hear environmentalists get on the bandwagon and make statements that could have quite a negative impact on us.'

Effluent is undoubtedly a major hurdle to dairying becoming an environmentally sustainable industry. Waterways and groundwater are polluted by the nitrate in cattle urine and nitrogen-based fertilisers, and studies suggest that half of this country's greenhouse emissions are caused by methane from farm animals and nitrous oxide from soils and fertilisers.

The seriousness of the problem can be gauged from the Dairying and Clean Streams Accord, signed between Fonterra, the Ministry of Agriculture and Forestry, the Ministry for the Environment and a number of regional councils in 2003. Under the accord, which has the backing of Federated Farmers, streams, rivers, lakes and wetlands must be fenced off, farm effluent appropriately treated and discharged, and nutrient run-off minimised. Environment Waikato is now monitoring compliance from the air.

Steve has already seen benefits from the new measures. Recycling water from the farm's effluent drain, for example, has put less pressure on fresh water, which is pumped from 100 m deep bores.

Research is now being directed at nitrate inhibitors to increase plant uptake of nitrogen compounds and reduce nitrate run-off. In the near future, however, it is those practical measures, and the new ways of thinking they encourage, that will have the greatest impact on agricultural pollution. I was aware of the powerful effect an environmentally attuned mindset could have after visiting Southlander Warrick Day in the Taringatura Hills north of Winton.

> Recycling water from the farm's effluent drain, for example, has put less pressure on fresh water, which is pumped from 100-m-deep bores.

Warrick's 820-ha farm was called Omokomaru ('Place of the Sheltered Lizard') and the name hinted at its ecological focus. Thoughtful, motivated and energetic, Warrick had turned much traditional farm thinking on its head and in the process received accolades and, with reluctance, conservation awards for his work.

Before making changes to any part of the property, he has been in the habit of studying it for a year or more to assess the environmental consequences. One surprising outcome was his decision to let the gorse alone in some areas.

'I don't say things against it. It is holding New Zealand together,' Warrick told me. He had been through it 'on hands and knees' and found a catalogue of emerging native plants — wineberry, kahikatea, whiteywood, broadleaf coprosma, cabbage trees, pittosporum — their seeds all carried by birds from native bush a hillside away and sheltered while growing by the nursery crop of gorse.

Warrick encouraged the same process of regeneration on stream banks and in economically worthless gullies, fencing them off and giving nature a free reign. Showing me a sparkling stream, protected by some of the 3.5 km of fencing that screened Omokomaru's waterways, Warrick spoke of the land's ability to 'heal itself' — an expression that came up often in his conversation. Now mayflies and freshwater crays proliferated, and the riverside plants sheltered fish and acted as a corridor for wildlife in a way no grazed bank ever did.

> Now mayflies and freshwater crays proliferated, and the riverside plants sheltered fish and acted as a corridor for wildlife in a way no grazed bank ever did.

I was reminded of an experiment by that proto-conservationist sheep farmer Herbert Guthrie-Smith on his Tutira station back in 1882. To test his belief that damage to the land could be reversed simply by protecting it from fire and cropping, he began noting the changes on an unused 10 ha hillside at his back door. Later he fenced off what had come to be called the 'Hanger', and until shortly before his death in 1940 he faithfully recorded its processes of renewal. No seeds were sown, no saplings planted, no unwanted aliens grubbed out or attacked with herbicide. Yet, seemingly to a preordained timetable, bracken gave way to manuka, which in turn yielded to softwoods and finally to totara, matai, rimu and other hardwoods. As if to humour the old man, tui, bellbirds, native pigeons and other long-absent birds returned and bred. Today, the Hanger is a well-known and much-studied piece of earth, and a fitting memorial to a far-sighted pioneer.

Nor were the aesthetics of living landscapes lost on succeeding generations of New Zealanders, though they may have been surrendered temporarily to utility. Almost half a century after the Hanger experiment began, but while Guthrie-Smith still lived, the journalist Alan Mulgan made a rhapsodic pilgrimage 'home' to England.

'The hedges that pattern the countryside, instead of fences, would by themselves fill the colonial with delight,' he wrote of the Old Country. 'A hedge has a "soul", even though in my country it may harbour deadly orchard disease and have to be rooted out, but a wire fence is a mere barrier, ugly and inanimate. When it is barbed it is venomous into the bargain.'

In 1927, when Mulgan wrote those words, New Zealand was in transition. The forest had been burned and cleared and the bare earth sown in grass, but little had been done to soften the new landscape or to restore it to ecological health. There was truth in the embittered quip that the country's coat of arms should be emblazoned with crossed axes and a box of matches.

In the fight to subdue the new land the early settlers had fought a little too hard. Restorative balance was needed, and it was a long time coming. A consequence was apparent when Cyclone Bola tore through Hawke's Bay and East Cape in March 1988, laying trees flat and carving off topsoil by the tonne. Some 1765 farmers were affected, with much of the damage occurring in steep hill country where little or no soil conservation work had been done, or on the 3600 ha of land covered by floodwaters. Horticultural and farming losses from the three-day storm were put at $90 million and the scale of the catastrophe galvanised farmers, politicians and conservationists into action. An enquiry set up after the cyclone recommended various flood-control measures, including soil conservation, river control and land-use planning, which were to be applied nationwide to minimise future flooding. A trust, the Landcare Foundation, was also set up to plant native trees on eroded hills and teach environmental lessons.

Mulgan's 'venomous' barbed wire is now a thing of the past — I have yet to meet a farmer with kind words for it — and lighter, more versatile, electric fences are increasingly replacing the non-barbed fences that remain. However, Warrick's practice of using native trees as shelter belts is yet to catch on. It should, he says. Native tree lanes filter the wind but, unlike dense shelter belts, allow good grass growth right up to the trees.

Another of Warrick's idiosyncrasies is his championing of red and silver tussock. The tussock that fills Omokomaru's gullies would be trashed by sheep in a matter of hours in winter if the intensive grazing was not properly managed, his wife Wendy told me.

'When we were overseas once, the farm manager we had employed left the sheep too long on strip grazing during rain.' She indicated a field. 'It took three years for that pasture to recover.'

While tussock wins little favour among farmers, Warrick told me that it had conservation value in terms of water retention and soil stabilisation. A big tussock could hold a surprising amount of water — up to 40 litres. 'They are almost as good as a forest, really.'

On my arrival, Warrick had taken me out the back door and pointed to a distant russet-coloured patch. It was 16 ha of red tussock, the last remnant in the Taringatura Hills. Over the horizon lay an undulating expanse of exotic plantation forest. 'It would be a sad day to see the tussock go,' Warrick said.

Not that he was averse to plantation forest. He had planted 95 ha of exotics over the past 11 years, including pines, red cedar, birch, alder, redwoods and Douglas fir. There were also decorative beeches around the waterways, which turned Omokomaru into something of a guilty pleasure. The impulse must have been strong in working hours to take out the picnic hamper rather than the winter feed. Nevertheless the plantings were practical, Warrick assured me. In addition to fostering a pleasant work environment, they created a microclimate that was favourable to grass growth.

The trees also helped form a flight corridor for the South Island tomtit, a diminutive forest bird that was 'suffering', to use Warrick's phrase, on the Southland Plains. The farm's elevated backblocks offered fine views over those same plains and out toward Foveaux Strait and Stewart Island, then in a sweeping arc toward Fiordland, the Takitimu Mountains and the Southern Alps. It was also possible to see, closer at hand, some of the farm's 40 ponds — the biggest of them alive with mallards, paradise shelduck, scaup, grey teal, Canada geese and shoveler.

Warrick eyed the panorama with satisfaction. 'I just love property I can flood,' he said.

Some 215 ha — a quarter of the farm — was either in wetlands, undeveloped or covenanted under the Queen Elizabeth II Trust. Before I left Omokomaru, Warrick led me up a hillside to where one of those covenanted areas, a rock outcrop, had been fenced off. We climbed the wires and made our way through unchecked grass and scrub to the rocks, which were etched in shadow. Peering into a crevice, Warrick straightened a finger. There, on a ledge, barely discernible in the dimness, was the slender tail of a gecko, the namesake of this place.

As I turned to leave, I marvelled that such an obvious tenderness toward such things could dovetail with the management of 4300 breeding ewes, 1800 hoggets and 150 or so cows and trading cattle.

In rolling Wairarapa sheep country, south of Carterton, I got another glimpse of the future of New Zealand farming. There, Mark and Susannah Guscott had taken the helm of Glen Eden, the 644 ha family farm, and with a combination of enterprise, youthful determination and 21st-century know-how were taking on the bewildering challenges of modern agriculture.

It was late evening when I turned off a country road lined with former rehab farms — the reward for war service — and up the Guscotts' long, metalled drive. Mark had telephoned earlier to excuse himself for the night — he had belatedly remembered his first wedding anniversary, the forgoing of which would threaten domestic equilibrium. I drove past what I took to be Mark and Suzanna's house and on to the old homestead which was headquarters, still, to Mark's parents, Phil and Jo.

Within minutes of my arrival headlights dazzled in the drive behind me. A car pulled up and Phil Guscott swung himself out with the habitual energy and purposefulness of a big man with much to do. Once inside, Jo filled the kettle and Phil set down a suitcase — he was back this very hour from a stint of farm consulting across the Strait — and gave some attention to conveying the accumulated experience of his 30-plus years of farming life.

There was a generous history to the place, going back six generations. The founder, John Milsome Jury, arrived in New Zealand aboard the whaling barque *Thetis* and jumped ship in the Bay of Islands. After a series of adventures which read like a picaresque novel, including falling in love with Te Aitu-o-te-rangi, the daughter of Maori chief Te Whatahoronui, and being captured by Te Rauparaha, he fetched up in the Wairarapa. After steadfastly refusing for many years to buy land, which he believed Maori shouldn't part with, he relented and in 1854 acquired 325 ha of flax and fern, which formed the heart of modern Glen Eden. The old family name is preserved in the nearby peak, South Jury, and a cousin has title to the original homestead, Glendower, an imposing two-storeyed residence of hand-sawn heart totara set amid trees on a nearby hillside.

Mischief was inadvertently done when Phil's grandfather took it upon himself to cut the original property into three long, thin strips for the benefit of his heirs. Within a generation they had become uneconomic. How the old gentleman thought such a device could have even an outside chance of working is unclear. What is certain is that it took all the industry and financial acumen Phil possessed to reassemble the severed parts that in time, inevitably, had slid through the family's fingers.

It was at this point in his story that Phil's realism, his resolute hard-headedness, became apparent. The future possibility of buying the old home, with its tennis court and secluded, overgrown swimming pool, its childhood memories and family resonances, had a strong emotional appeal that would have lured many a person. Phil resisted the siren call, though, and instead bought a 179 ha farm down the road as an investment for his two daughters, Georgina and Charlotte. The new property was currently grazing ground for the farm's 'B' flock and attractive black-faced terminal-sire Suffolks. It also contained a parcel of land that would do nicely as the site for Phil and Jo's retirement house, once Mark came knocking on the door of their present home.

It was all part of that fundamental, and often neglected, department of agricultural science called succession planning. These days it accounted for a significant chunk of Phil's consulting workload — his trip to Banks Peninsula had been about precisely that.

Phil set out his four principles of effective succession planning. First, the family must be kept together. In other words, the logic of the new arrangements needed to be explained and understood to avoid bitterness and resentment and even the kindling of a family feud that could remain alight for years. Then the plan should provide the retiring generation with a place to live and an income. It must also create a viable business for the next farming generation. Lastly, there must be a financial plan for non-farming family members. 'No farming family can duck those issues if it wants to remain a family,' Phil said.

His own journey toward consulting was salutary. The 1950s and 1960s were fine decades to be a farmer, he said, but by the time he arrived in 1978 the economic noose was already tightening. The uncompromising economic reform introduced by finance minister Roger Douglas spun Phil and countless other farmers around in its wake. Faced with either selling the farm or working away from it, he became a farm management consultant and rural valuer.

A habitual early riser — up at 5.30 'since I don't know when' — Phil found time in 1987, around the edges of his existing commitments, to start a meat export business. 'The idea was the easy part. Thereafter it got hard,' he said. 'We delivered our first meat to a San Francisco supermarket in the boot of a rented car.'

Based in Hawke's Bay and now employing 30 staff, the company sea-freights chilled meat to the United States. The early years were not easy. Nor did profitability come quickly.

'We were no different from a lot of farmers. Everyone was pushed to the wall back then,' Phil said. The export business began, he explained, because he was 'hacked off' at the pittance being paid for lamb.

It was not widely appreciated what the sudden removal of subsidies and all the other structural changes of the 1980s did to the rural sector, he said. Besides the impact on farmers, there was the flow-on effect to topdressing pilots, transport companies and all the other service industries. 'In my first year as a consultant, I spent most of my time going to creditors' meetings. But that's life. That's what happens.'

By comparison, the past five years had been 'bloody brilliant'. Nevertheless, wool was fetching half what it had in 1987 and in the 2006 year alone costs were up five per cent and farm income down 20 per cent. 'If the government doesn't control interest rates it will kill us faster than anything because we can't put our prices up.' Phil quoted the old truism that a farmer sold produce wholesale, bought supplies retail and paid the transport charges both ways. 'It is a simple equation that has been around for the past half-million years and it will never change.'

One response to the equation, as I knew, was to court economies of scale. Another was to improve efficiency by harnessing technology. For Phil that did not mean nursing a love of machinery for its own sake — what he called 'heavy metal disease'. That, he said severely, was the death knell to a farmer. No, what he had in mind was modern equipment that enabled farmers to work smarter. Computer-controlled irrigation, for instance.

The next morning he took me out to the back boundary of Glen Eden to see an enormous circular paddock that had been created to fit with geometrical precision into the serpentine bend of the Ruamahanga River. It was home to one of the Guscotts' most effective innovations, a drought-proofing centre-pivot irrigator with a 600 m long arm capable of delivering 60 litres a second, or 10 mm of water in a single 48-hour pass. As a result of using the irrigator over the summer months, pasture growth had climbed from 10 to 12 kg/ha of dry matter a day to 100 kg/ha.

Ironically, the irrigator itself was under constant threat from floods — what they called 'Pivot Flat' had been awash eight times the year before last. It took five hours to get the irrigator arm out of danger, facing away from the river. The regional council ran a system of automated rain gauges in the hills which gave an early warning, though with five tributaries to account for, accuracy was difficult. 'If the phone goes in the middle of the night you know why. It always happens in the middle of the night. If we have a four metre warning we will rip out there and do something about it.'

Installation of the irrigator led to soul-searching as to the best use for Pivot

Flat. Cropping was good out the back, said Phil, 'but the old river removes your enthusiasm for it.' Then there was dairying. The numbers stacked up for cows — $5000 income per hectare, compared with $2000 for sheep and beef — but if you are not dairy-inclined then you are not. In any event, irrigation allowed the Guscotts to become lamb finishers themselves instead of selling live lambs to other farmers. It was a worthwhile benefit as lamb income represented around half of farm revenue, the rest derived from ewes, wool and beef.

Climbing the rutted road from the flat to the hilly backbone of the property, Phil stopped and turned to enjoy the prospect. 'In Dad's day they had a mob of 3000 sheep down there,' he said. 'Ten men and 50 dogs were needed to move them to the front of the farm for shearing. They spent the first hour sorting out the dog fights, because they all wanted to be top dog. Then it took until mid-morning to go half a kilometre to this elbow.'

From this comfortable distance, it was an entertaining thought. To neighbours lending a hand at the time the amusement no doubt would have been fugitive. The lanes were not fenced in those days and under a warm sun it would have been a trying business. Then again, perhaps they exercised that enviable trait I had noticed among New Zealand farmers — humour in the face of adversity.

Now the farm has more compartments than the Titanic, along with 500 or so gates to swing and chain, and anarchic freedom has given way to the despotism of order.

During the First World War, the New Zealand army had introduced its own zest to farm routines by periodically training artillery on South Jury and environs from positions at Papawai and lobbing shells across the intervening five or more kilometres. In a sporting gesture, Phil's grandfather was given sufficient warning to move his stock, but evidence of the reality of the threat is still being unearthed on the farm. Phil has a jar of grapeshot — small lead balls the size of a thumbnail — found on farm tracks after rain.

'In Dad's day they had a mob of 3000 sheep down there,' he said. 'Ten men and 50 dogs were needed to move them to the front of the farm for shearing. They spent the first hour sorting out the dog fights . . .'

It could not have been easy for Phil to practise what he pronounced when it came to succession. Any vocation followed with passion for half a lifetime is hard to walk away from. When it involves an intimate attachment to the land, the letting go must be even harder, especially when there was a new generation stepping up to the plate, keen to experiment, to add its own ideas to the mix, and to happily get wrong what had once been right.

At least the consulting proved a distraction, and in the weekends there was always fencing and welding. 'Mending fences is my recreation now,' said Phil. 'After 15-hour work days, I need the solitude.'

Phil was also fortunate in the quality of Glen Eden's next generation. But before spending time with Mark, who had inherited his father's boundless energy and farming savvy — he was a finalist in the regional Young Farmer of the Year award — a visit to the shearing shed was called for.

The shed was alive with the clatter and buzz of shears when I got there. After a three-week break following the main shear, the gang was hard at it. Shaun, the presser, was straining to compress a bail, and the rousies were gathering the fleeces with all the attentiveness of a first day back on the job. It would take just over two days for the three shearers to put the 1400 sheep, all two-tooth ewes, through the stalls. The ewes were going to the ram in a month and this late shear was more in the way of an animal-health exercise.

'The price of shearing has gone up by a dollar a sheep in the past year, so everyone is exercising their minds about how to do it less.'

With income from wool barely covering the cost of getting it off the sheep's back, shearing was increasingly done as a part of stock management. 'The price of shearing has gone up by a dollar a sheep in the past year, so everyone is exercising their minds about how to do it less,' Phil had told me.

Glen Eden was the gang's biggest contract. They had clipped 6500 sheep here in January, but the winter shear, which used a big comb to leave a layer of wool on the skin, was now starting to become common in the Wairarapa, as it had always had been in the South Island, where it is known as the pre-lamb shear. Gangs even resorted to blade shears in the high country around Blenheim to leave more wool on the animal.

Gordon Bassett, the ganger, had definite views about where shearing was heading. Apart from messing with his shearing calendar, sheep farming was causing havoc in the stalls with its new breeds. 'Composites are no good for us. They are more skittish and fiery,' he told me. A rest had been called, and Gordon sat on the wool-littered shearing boards, his back against a wall. 'A Romney would just sit there and let us do the deed.'

Still, at least the sheds on New Zealand farms were functional. In his 34-year career Gordon had seen a good deal of the world's approach to shearing, and apart from the competition across the Tasman there was not much to say in its favour. Germany, surprisingly, was primitive. 'Sometimes just shearing gear hung up under a tree. Other times, in a barn.' England offered a slight improvement: Gordon had worked from a purpose-built trailer set up in a field. 'Just fire up the generator and away you go.'

The radio was full of the Golden Shears competition in Masterton when I left the shed, stepping past a spray of sparks from a comb being sharpened on a wheel and out in the direction of Mark Guscott.

I caught up with him down on Pivot Flat, fiddling with the door springs of an automatic weighing machine. At his back ram lambs pressed impatiently in the race. The idea was to weigh and record the animals one at a time, after which the door would be triggered to one of two holding pens, depending on the size of the animal. 'We're trying to get this down to a — settle down, Duke!' Mark rebuked a dog that was hurrying up the sheep regardless. '. . . down to a one-man operation,' he said, persevering grimly with the spring.

I raised the subject of heavy metal disease. 'I have to say, I'm with Dad on that one,' Mark said. Nevertheless, it was clear that he was not averse to working smarter. Once the lambs had been weighed, they were identified with a blue spray. 'I'm hoping someone will invent a machine for that,' he said. 'There is a field day in Feilding in about a month. That's the place to find that sort of stuff.'

Not the only place, though. The farming world is awash with information these days. Mark had subscriptions to five industry publications totalling 14 issues a month, the pooled knowledge of a regular 25-strong farmers' discussion group, and access to that great postmodern grab bag, the internet.

Already useful for paying invoices, the internet had huge potential. A webcam could be mounted on the irrigator or on gates as part of an automated feeding system. In Australia, farmers were accessing satellite readings of grass growth. 'We have 150 paddocks here and I usually check the levels with my "eyeometer". A satellite would make the job easier.'

Hearing a sheep man talk of extraterrestrial tools was exhilarating, but then that was the curve farming had been on from the time the first sheep came ashore. We are entering an age of freeze-dried, biologically active proteins and touch-sensitive fabrics, of cloning and transgenics, nutraceuticals and animal genomics.

Where such things contribute directly to improved productivity, pragmatic New Zealand farmers will use them. There were common threads in that great unbroken chain of enterprise stretching from Samuel Butler and Charles Suisted to Lady Barker and Herbert Guthrie-Smith and on to the Guscotts and all the other motivated and resourceful men and women I had met in my travels across two islands.

'The first rule of business is to stay in business,' Mark had said. Attend to the needs of the farm, in other words — to its physical health and economic wellbeing — and the farm will deliver your future.

Steve and Julie Nelson's dairy farm in the rich Waikato milks a mixed herd of 650 Friesian and Jersey cows. This herringbone milkshed is soon to be replaced by a more efficient 50 unit rotary shed.

A measured feed of turnips helps maintain the high milk production of these Friesians. A favourite among dairy farmers, Friesians produce large offspring, ensuring a good return on calves.

259 *the* Future

↑ This hayrake now has a merely decorative use as a result of the continual evolution of farm technology.

↓ Cows gather round the bail feeder — one of the daily routines that measure out life on a dairy farm.

→ On the Nelson's leased runoff farm the expense of plastic-wrapped silage is outweighed by its ease of use. The home farm still uses the traditional stack and silage wagon.

263 *the* Future

265 *the* Future

Taking on the afternoon shift, Steve Nelson attends to his Jersey cows, known for their ease of handling and high milk solids.

A shared family meal is a chance to catch up on the morning's news. For Steve the demanding milking schedule is largely a thing of the past now that his son Daniel (centre) runs the farm.

A hillside of tussock, valued for its water retention and land stabilising qualities, overlooks the Southland Plains near Winton, where award-winning Omokomaru farm is a showcase for ecological innovation.

269 the Future

Omokomaru's owner Warrick Day is transforming his 820 ha property by fencing off and replanting waterways, building ponds and introducing native shelter belts. A quarter of the farm is now either in wetlands, undeveloped or under protective covenants.

Romney sheep are penned for the main shear at Omokomaru. Warrick Day runs 4300 breeding ewes and 1800 hoggets as well as some 150 cows and trading cattle.

Shearing instructor Brendon Potae passes on his knowledge to young English shearers. Gangs dislike working with the newer composites, raised for meat, which can be difficult to handle in the shed.

275 the Future

277 the Future

With poor international prices for wool, shearing is increasingly viewed as little more than a stock management exercise on many New Zealand farms.

Grazing Romney sheep catch the early light at Glen Eden farm in the Wairarapa. Family head Phil Guscott worked hard to reassemble the 644 ha farm after an earlier generation allowed much of it to pass out of the family.

→ Susannah and sixth-generation farmer Mark Guscott now run Glen Eden. Rather than dividing the farm, additional land has been bought nearby for his sisters. To keep farms viable succession planning is vital.

↓ Stock is trucked from Glen Eden to the Guscotts' nearby second farm. Mark specialises in producing meat for the North American market.

283 the Future

⬆ Mark Guscott checks the quality and growth of grass in a newly-drilled paddock at Glen Eden.

➡ Information technology is enabling farmers to capture and analyse vast amounts of data in order to micromanage stock and pasture. Mark recently signed up for a broadband internet connection and talks of hooking up a remote camera to monitor the distant irrigator.

⬇ The farm's original woolshed, now listed by the Historic Places Trust, stands as an eloquent monument to the pioneers who laid the foundations of New Zealand's agricultural prosperity.

287 *the* Future

Vaughan Yarwood

Vaughan Yarwood is the author of *The History Makers: Adventures in New Zealand Biography* and *Between Coasts: from Kaipara to Kawau*. A former editor of *Management* magazine and associate editor of *New Zealand Geographic*, he has written widely for New Zealand and international publications.

He recently edited an electronic web-delivered edition of his earlier *Covering Asia: A New Zealand journalist's handbook* and contributed introductory chapters on New Zealand's geography and environment for the latest editions of the Lonely Planet guides *New Zealand* and *Australia & New Zealand on a Shoestring*.

International assignments have taken him to many countries in Europe, Asia and the Pacific.

Arno Gasteiger

Arno Gasteiger was born in Austria and moved to New Zealand to live and work in 1988. He specializes in location photography for editorial and commercial clients. His passion for photography has earned him several major awards and assignments from some of the leading magazines in the USA and Europe.

He is the author of six books. *Central*, a large photographic book on Central Otago, won the illustrative category at the 2004 Montana Book Awards.

His photographs are widely exhibited and many of his works are held in collections in Europe, New Zealand, Australia and the USA.

www.arno.co.nz